THE DEATH OF Reliability

THE DEATH OF Reliability

Is it Too Late
to Resurrect the Last,
True Competitive Advantage

Dr. Nathan C. Wright

Industrial Press, Inc.

Industrial Press, Inc.
32 Haviland Street, Suite 3
South Norwalk, CT 06854
Phone: 203-956-5593
Toll-Free in USA: 888-528-7852
Fax: 203-354-9391
Email: info@industrialpress.com

Author: Dr. Nathan C. Wright
Title: The Death of Reliability: Is it Too Late to Resurrect the Last, True Competitive Advantage, First Edition
Library of Congress Control Number: 2017947357

Printed in the United States of America.

ISBN (print): 978-0-8311-3622-2
ISBN (ePDF): 978-0-8311-9461-1
ISBN (ePUB): 978-0-8311-9462-8
ISBN (eMobi): 978-0-8311-9463-5

Editorial Director: Judy Bass
Copy Editor: Janice Gold
Interior Text and Cover Designer: Janet Romano-Murray

industrialpress.com
ebooks.industrialpress.com

Dedication

*For my loving and supportive wife, Lisa,
and my four wonderful children,
Joshua, Stephanie, Troy, and Natalee,
who provide unending inspiration.*

Table of Contents

Acknowledgments

This work would not have been possible without the support of the maintenance and reliability professionals I have had the pleasure to work alongside over the past thirty-five years, particularly Bill Cole and Larry Wiskirchen. I am especially indebted to Steve Holmes, Duane Aldueso, and Robert Hansen who have shaped my career and who worked actively to provide me encouragement to pursue my goals.

I am grateful to all of those with whom I have had the pleasure to work with during this project. Each of the members of my publishing team has provided me extensive personal and professional guidance and taught me a great deal about turning my life's practices into a volume. I would especially like to thank Judy Bass, the Editorial Director, Janice Gold, my copy editor, and Janet Romano-Murray, my cover and text designer. As my publishing team, you will be forever linked to my family. I would also like to thank my brother and sister, Lester and Tracy, and my sons, Joshua and Troy, for reviewing my work to ensure I am making sense and adding value to the body of work.

Nobody has been more important to me in the pursuit of this project than the members of my family. I would like to thank my parents; whose love and guidance are with me in whatever I pursue. They are the ultimate role models. Most importantly, I wish to thank my loving and supportive wife, Lisa, and my four wonderful children, Joshua, Stephanie, Troy, and Natalee, who provide unending inspiration.

About the Author

Why do I believe I have the answer to the question of how to make reliability a competitive advantage? I would like to give you a little background on me to put into context why I feel I can talk on this subject. I am a fourth-generation reliability person. My great-grandfather, grandfather, and father all worked for the Phelps Dodge Corporation. I have lived firsthand the effects of leadership in our industry. While some folks network with a few folks to share experience and knowledge, I lived it daily. Not only was my father a maintenance manager, but his brothers, as well as my mother's brothers. My brother, cousins, nephews, and sister-in-law are all skilled tradespeople. My son is a fifth-generation reliability guy along with his cousins, so my network is extensive and spans many industries. While writing this book, I asked my son to read the draft and comment. As he and I sat discussing his review, he stated that most of what he read he has heard discussed at every family get together. I do not have to go to a seminar to network; all I must do is show up to dinner. This was confirmed by my son during our discussion, and it has been the same for him.

MY EDUCATION

From an education standpoint, after I graduated from Morenci High School I attended the University of Arizona and studied electrical engineering. Upon my father's death, I enlisted in the U.S. Navy to continue my studies in electrical engineering and surprisingly, leadership. The Navy fueled my desire to understand how leadership affects everything else. The one thing the Navy taught me was that with advancement to positions of increasing leadership, there needs to be significant development. Now let's fast forward to all my other promotions outside the Navy. There was little, and by that I mean no, leadership development. I was promoted because I was good at the position I was in and it was assumed that I would figure it out as I went along. Without development, there is a lot of risk to the success of leaders and their effects on their teams. As I worked to better understand the aspects of leadership, I pursued my bachelor's degree because it is the latest requirement and belief that a reliability leader needs a degree. I attained my MBA so I could better understand the financial aspect of my job. It is no longer enough to know the right thing to do because companies today focus solely on the bottom line and not return on investment. To be successful as a reliability leader, I had to be able to talk in the language of today's leaders: finance and budgets. Because the C-suite leaders are no longer people who have risen through the ranks, many corporate leaders lack real-world experience and an understanding of how to run the plant. They are lawyers and accountants focused on short-term objectives and not the long-term success of the company. To gain their buy-in I had to learn to speak their language. Finally, I completed my doctorate to enhance my ability to lead. With the ever-increasing pressure to "do the wrong thing," leaders need to be able to motivate their teams despite poor corporate leadership. In order to deliver the results I desired with respect to reliability, I had to understand how to change the organization's culture. The ability to know what to do is complemented by knowing how to change the culture that opposes it.

MY CERTIFICATIONS AND LICENSES

In addition to my education, I have obtained several certifications and licenses to further enhance my ability to deliver reliability to any organization I work within. I obtained my certified maintenance and reliability professional (CMRP) in order to communicate my 35 years of maintenance and reliability experience. This lets others know that besides my experience, I do carry the only ANSI certification in reliability, giving credibility to my experience. I also hold a project management professional's (PMP) license. Moving an organization from run-to-failure maintenance to proactive reliability is a series of projects and I wanted to ensure I could be effective at leading and managing these projects. Delivering reliability relies on delivering projects on time and within budget. Each successful project builds on the confidence necessary to make these cultural changes. The next certification I pursued was the International Council on Machinery Lubrication's machinery lubrication technician level one certification. With lubrication being 70% of the root cause of unreliability, it made sense to acquire a certification focusing in this area. As you can see, every degree, certification, or license was achieved to make me a better reliability professional.

WORK HISTORY

After leaving the Navy, I worked my way up the ranks, first at Phelps Dodge Tyrone then P&H Mining, followed by Northrop Grumman, ASARCO, Nyrstar, Empire Southwest, Kraft Foods, and Continental Building Products. I have been a laborer, apprentice, journeyman, front line supervisor, planner, superintendent, manager, and corporate officer, and the most important part of each one of these positions is leadership.

That is why I have written this book. It is my hope that we can turn the tide of organizations looking for shortcuts and immediate gratification, which has eroded the competitive advantage once held in the U.S.: Qualified Reliability Professionals.

Testimonials

I have had the privilege of working with Nathan at 3 different companies. All were in mining and wanted to have a reliability based maintenance program. Nathan is a back to basic type of leader. He knows that to build a good program you start at the bottom and build a good base. That starts with employees. He says you need people that have the skills whether it's planners, journeyman, and apprentices. Nathan also believed that you must have buy-in from operations and management. Without it you cannot make things happen. I have personally seen his program work and decrease the cost of maintenance and increase safety and retention of workers. If I had the chance, I would work on another project with Nathan.

Bill Cole
Electrical Supervisor
Kinross Gold Corporation
Round Mountain, Nevada

For over 25 years I have had the privilege of observing and working with hundreds of exceptional leaders and specialists in a broad spectrum of business disciplines. My association with Dr. Nathan Wright has been one of the highlights of that experience.

Dr. Wright has a unique grasp of the realities of the business world and the often-perplexing challenges associated with focusing and channeling the energies of leaders, managers and workers within an organization. His insights come from his extensive background in manufacturing, mining, and consumer products, giving him the ability to see the dynamic balance between the policy and process issues of an organization and the people issues that make or break implementation and execution.

As a teacher of leadership skills, he brings the hands-on experience of having turned around ineffective teams by developing both the skill set and the mindset of leadership in those he has coached and mentored. He has a no-nonsense approach to training and development, which stems from a keen awareness of what people need to know and do, and what they don't need to do, in order to achieve the outcomes demanded in today's global economy.

Dr. Dennis Deaton
Co-founder and CEO
Quma Learning Systems, Inc. Mesa, AZ

Author, The Book on Mind Management, Ownership Spirit: The One Grand Key that Changes Everything Else

In his new book: The Death of Reliability: Is it Too Late to Resurrect the Last, True Competitive Advantage, Dr. Nathan Wright takes innovative thinking to a new level. His approach to reliability could be the missing piece needed to complete the maintenance puzzle and bring about the increase in reliability that today's engineers and managers are looking for, but unable to find.

If Dr. Wright is correct in his analysis of the challenges manufacturing facilities face regarding reliability, then industry is indeed in a very dire position and needs to act quickly before it is buried beneath its own mistakes.

Before Nathan led the maintenance team at the plant where I was working, maintenance was mostly reactive and poorly organized. Within a few weeks, he had begun to introduce the ideas he expands upon in this book which brought the plant uptimes which had not previously been thought possible. It was clear that his rich background in maintenance, reliability and management was a huge advantage to any organization.

I believe Dr. Wright is in fact correct in his analysis, and that his book will be a huge advantage to your organization as well.

Larry Wiskirchen
Reliability Professional,
Vibration analyst, ISO Cat. IV

I have known Nathan professionally for more than ten years. He has a unique combination of journeyman skills and advanced education that is extremely rare and valuable.

I completely agree with Nathan's views on the exceptional value great reliability can bring to the bottom line. Many of today's business leaders don't fully understand the value that great, consistent reliability can bring to the success of a business. And they certainly do not properly value the experience and talent a skilled craftsperson brings in gaining reliability. Journeyman training programs are being replaced with abbreviated technical programs to speed employees into the field where they are unprepared to face some of the complex problems they will be expected to resolve in their jobs.

How we encourage young people to go down this path, which takes real commitment, hard work, and dedication starts with parents, teachers and businesses. If the growing gap and deficit in truly skilled reliability leadership continues as Nathan suggests, there will be a lot of business pain along this long road. But today's leaders can get a jump start by supporting the certified, comprehensive journeyman training programs for their developing employees.

In today's business world, we see a huge number of senior executives who have excellent backgrounds in legal, accounting, financial, etc., but lack a basic understanding and knowledge on what truly drives high performance in their business. Consultants are overused as a quick fix, resulting in huge expenses and few lasting improvements in many cases. Nathan is right in that it is this disconnection between many senior executives and "the business" that leads to a lack of support for nurturing and developing strong craft skill sets in the workforce.

I especially liked Nathan's write-up on Transformational Leadership—empowering great people to perform and reach their full potential. They motivate and support their leadership teams, are positive thinkers, create inspirational vision that drives motivation into their teams. The opposite, Industrial Leaders (transactional), are all about position, policy, and micromanagement—the death of great employees.

Leaders should recognize the value in skilled craftspeople, and make sure they understand the connection between hands-on experience, matched with strong leadership, and the value great reliability brings to a business. I fully agree with many of Nathan's positions and see a huge value for leaders and business managers to invest in this book.

Steve Holmes
Chief Operating Officer
KGHM International, a subsidiary of KGHM Polska Miedz

Preface

Are we facing the death of real reliability? The simple definition of *death* in Webster's Dictionary is "the permanent end of something that is not alive: the ruin or destruction of something." The sad truth is that this is exactly where we are headed with respect to reliability. The reason I believe this is true is the loss of real reliability leadership. Without qualified leadership, there can be no real reliability and as such companies are losing out on the one true competitive advantage available to them today. I have repeatedly heard that the loss of skilled craftsmen stems from kids today not wanting to work hard. I believe that this is a small portion of the issue. There is a lot more behind the views of kids today than meets the eye. The erosion of individuals seeking skilled trades as a career starts at home. Parents fill their children's heads with the belief that everyone should go to college. This cannot be further from the truth. The same message, that college is the only answer, is reinforced by teachers and counselors. The problem with these teachers and counselors is that they have never worked in a skilled trade and the majority hold non-technical degrees and know nothing of the trades industry. This anti-skilled trade attitude is prominent throughout our government. The ill-conceived "No Child Left Behind" program further damaged the noble professions in favor of non-technical educations that provide few options for the holders. When compared to the income of a college graduate, especially a non-technical graduate, a skilled tradesperson can make two or more times their salary. This fact has been presented to the last three presidents by Mike Rowe, a television personality best known for his show D*irty Jobs*. Like me, he has gathered facts supporting this position. While we have different approaches to this message, we stand together in our efforts to open the eyes of anyone who will listen.

This gave me pause and made me want to write this book. After reviewing my background and experience, you will understand the context from which I draw in the book that follows. Unlike most of the books written on this topic, this book is not the ravings of an engineer or consultant who is espousing theory and other information based on no hands-on, real-world experience. I have read many of these volumes and found no hands-on facts, but a lot of supposition about the what and the why but nothing on the how. Please do not take my statements here and in this book about the lack of qualifications by the majority of maintenance and reliability leaders as condemnation of the individuals. It is not a condemnation of the individuals but of the wave of poor decision making by senior leaders worldwide to shortcut the hard work required to develop a true reliability leader.

Another big driver of this issue is the lack of leadership at all levels of organizations that understands the competitive advantage of reliability or reliability itself. Instead, they pat themselves on the back for attempting to shortcut what it takes to achieve reliability. They "fudge" the numbers to make it appear that they are progressing in the right direction, but what they cannot do is cover up the unreliability that causes them to lose ground and increase costs. My father always told me "figures never lie, but liars can figure" and that is the "new" leadership style for those posing as reliability leaders today. Just thirty years ago, an organization invested time and money in their employees to develop craftspeople. It is the loss of these efforts that is driving the decline of craftspeople. The misguided shortcuts being employed by organizations to attempt to circumvent hard work and effort are eroding craft skills and many other one-time strengths of the United States. The belief that abbreviated training can replace apprenticeships and produce skilled craftspeople is completely wrong. This belief is driven by the loss of leadership in the senior positions in organizations. Specifically, the loss of real reliability professionals in key senior positions is decimating the skilled craftspeople ranks. Before you blame the kids of today for this loss, you need to look at the leaders of today: parents, teachers, government, plant managers, C-suite officers, and board members.

WHERE WE WERE

On December 7, 1941, we suffered a surprise attack on our Navy in Pearl Harbor, Hawaii. This crippled our naval forces, and it was the industrial might of the U.S. that bailed us out of what could have been our defeat. Admiral Isoroku Yamamoto, Fleet Admiral and Commander-in-Chief of the Japanese Imperial Navy, upon being told of the success of their attack on Pearl Harbor, stated, "I fear we have awakened a sleeping giant and filled him with a terrible resolve." His statement was the result of his experience of traveling in the U.S. and seeing the industrial might we possessed. This industrial might was rooted in the strength of our skilled tradespeople. He knew we could easily shift from making cars to tanks and planes. We could shift from making tractors, plows, and manufacturing equipment to weapons that would be superior to the Axis forces. With the current state of skilled trades in our nation today, we would not be as successful and I would say we might not even prevail if today's attacks were focused on military targets, not helpless civilians. The purpose of this publication is to bring attention to the crisis facing skilled trades and craftspeople. It identifies the root cause of the impending death of reliability if we do nothing. Edmund Burke said that "the only thing necessary for the triumph of evil is for good men to do nothing." It is my hope that I can reach enough people to incite thought and debate about this infection and the immediate need to begin CPR. Ultimately, with a lot of hard work and effort (rehabilitation), we can return the reliability industry back to good health, with good people doing something. The answer to the "what's in it for us" question is a competitive advantage.

When I say, "I" it is my hope that you realize that what I am conveying is what *I* have experienced. It is not meant to preach or tell you my approach is better. When I use "we," it is my intention to let you know that I am with you in spirit and it is meant to convey a shared experience.

Introduction

Competitive Advantage

After spending 35 years in the workforce, I know that the goal of all organizations is the pursuit of a competitive advantage. Webster's Dictionary defines a *competitive advantage* as "superiority gained by an organization when it can provide the same value as its competitors but at a lower price, or can charge higher prices by providing greater value through differentiation." Competitive advantage results from matching core competencies to the opportunities. The irony of this pursuit is that organizations are seeking to achieve this by working towards "best practices." There is no greater illustration of a herd mindset than that of best practices. I have always held that a so-called best practice ends the minute it is labeled as such. Best practices defend the status quo and restrict innovation by ensuring people/processes follow the same tactics. You cannot distinguish yourself by embracing sameness. The idea of best practices is basically the unification to the norm at your own peril. Smart leaders innovate past best practices and are always in search of next practices. If your decision to do something is driven by others doing it the same way, you are doing little more than yielding advantage and opportunity to those opponents more creative than you. Don't copy—create. If your organization is looking for the last remaining competitive advantage, I challenge you to read this introduction, and I know it will motivate you to finish the book.

THE BEGINNING

In the quest for a competitive advantage, the world has undergone many transitional efforts. One of these endeavors is the search for process improvement. There have been occurrences of process reasoning in manufacturing that date back to 1100 A.D. and the "Venice Arsenal." Henry Ford was the first person to incorporate the model production process. In 1913, he combined interchangeable parts with an assembly line and standard work to create what he called flow production. Ford's problem was his inability to provide variety. He is known to have said that "you can get the Model T in any color you want if it is black." Here is where Kiichrio Toyoda and Toyota discovered that they could provide variety and maintain continuity of their process flow while revisiting Ford's original approach, resulting in the Toyota Production System. This system changed the focus from the individual machines to the entire process and made it possible to obtain high variety, low costs, and rapid throughput to respond to customers desires. This system has been labeled a shining example of lean manufacturing. The basic rules of lean are to: identify the value desired by the customer, identify the value stream and eliminate waste, make the products flow through value-added steps, introduce a pull between all steps, and manage towards perfection. This system made Toyota an example of lean to the world. Its success in the implementation and use of lean led to Toyota's global dominance and stands as validation of the lean concept. The success of lean over the past several decades has created a huge demand for knowledge about lean thinking. There are thousands of papers, books, and media articles investigating the subject. As lean thinking spreads to every country, leaders rush to adapt the tools and principles beyond manufacturing. Lean consciousness is beginning to take root in all sectors, but the belief that anyone can implement any system successfully simply by attending a seminar or taking a few classes, or by having a college degree has prevented the organizations from achieving true greatness. Any true achievement and competitive advantage can only be gained by hard work and leadership.

LEAN VS. RELIABILITY

Lean manufacturing has it place in process improvements. My concern is the erroneous belief that lean can improve reliability. What started as a process-focused initiative has become a broad reaching fix-all that fails in its mission. The use of lean tools in reliability efforts has resulted in attempts by senior leaders to short cut the hard work required to achieve true reliability. Getting a group of people together to figure things out is necessary because few organizations have qualified people leading their efforts, so there is a need to have a meeting or several to groupthink every problem. Lean is another example of herd mentality or group thinking. A qualified reliability professional leading your team will know what the root cause of the problem is because of their knowledge and experience. Three things cause unreliability (improper lubrication, contamination, and improper installation,) and I do not need a cross functional team to identify them. Lean, like all other consultant driven initiatives, is a fancy sounding program that over promises and under delivers. I find myself laughing when senior leaders preach *Total Productive Maintenance* (TPM) as the pinnacle of maintenance, because it is such only in an unqualified reliability professional's mind. Again, this is proof of what I am saying in this book. There is no shortcut, quick fix, or wonder drug to fix reliability. It is as simple as hard work and qualified leadership. You will find yourself throughout this book either agreeing or disagreeing with me, but the bottom line is I challenge you to really look at why you disagree with me; it is probably because this hits too close to home.

There are no shortcuts or one-size-fits-all programs. Consultants cannot get you there. Shortcuts and consultants will devour your money and waste your time. Look within your organization for leaders and develop them to achieve results. If you do not have a reliability professional in your organization, then hire one and invest your time and money in success.

TRAINING VS. DEVELOPMENT

Another favorite tool of organizational leaders today in the hunt for competitive advantage is the misguided training efforts of their leaders. Organizations have spent billions each year attempting to train their leaders. They reference Jim Collins' book, *Good to Great,* in many of these training sessions. Several of these organizations will make reading this book mandatory in the hope that the information contained within will magically transform their leaders into those responsible for the success of the eleven companies in Jim's book. "Good is the enemy of great. And that is one of the key reasons why we have so few entities that become great. We don't have great schools, principally because we have good schools. We don't have a great government, principally because we have good government. Few people attain great lives, in large part because it is just so easy to settle for a good life." To give a business the best opportunity to succeed, leaders should be chosen for a role based on their natural ability to flourish in that role. After they are selected, leaders should be given opportunities to develop and grow in ways that will help them meet the requirements of their respective roles. This tailored approach diminishes the desire to provide ineffective, "one-size-fits-all" training and allows the leaders to concentrate on the areas that will make them and their organizations most successful.

If every organization is in hot pursuit of a competitive advantage, how do you find the real deal? Like I stated at the beginning, in my experience I have encountered all the different approaches in this endeavor. It is my goal in this book to share these experiences and assist you in your pursuit of competitive advantage and ultimately show you the untouched opportunity still available to every organization. Many organizations believe, falsely, that they have identified this opportunity but fail to truly understand what it is.

THE OPPORTUNITY

The opportunity I am speaking of is reliability. This opportunity is the result of true reliability, and more specifically, the leadership of reliability. Organizations need a reliability professional directing their efforts. A reliability professional has a lifetime of real-world experience, a journeyman's license, certifications, professional licenses, and a professional reputation of success. To develop and lead your reliability teams they must have worked as a combination of reliability person, planner, supervisor, reliability engineer, and reliability manager. In addition to this level of practical real-world experience, they need to possess an advanced education with finance, development, change, and leadership concentrations.

Reliability leadership is not a growth position for young engineers nor a position delegated to production managers. Look again to Jim Collins' book, *Good to Great,* in which he observes the necessity of having the right people on the bus. "Letting the wrong people hang around is unfair to all the right people, as they inevitably find themselves compensating for the inadequacies of the wrong people. Worse, it can drive away the best people. Strong performers are intrinsically motivated by performance, and when they see their efforts impeded by carrying extra weight, they eventually become frustrated." Consultants, engineers, and production personnel are not reliability professionals and are the wrong people to lead these efforts. This is not my isolated observation but one repeated throughout literature. Daniel Daly, author of *The Little Black Book of Reliability Management,* a mechanical engineer, professional engineer, and certified maintenance and reliability professional, states, "I began working as a mechanical engineer. The things I learned in college and the books I used were of limited value." He goes on to discuss how the experienced folks were the true teachers. He refers to them as the "old heads." The current propensity to fill reliability management positions with the wrong people has prevented organizations' transformation from good to great. The first step is to build the right bus and then get the right people on it.

Let's break this down a little further to help everyone understand what I am saying. When we compare a four-year engineering student to a four-year apprenticeship with the end goal of being a reliability professional, there is no "apples to apples" comparison. During a four-year apprenticeship, an individual will learn how to: troubleshoot electrical and mechanical systems, align these systems, identify when they are not functioning at their optimal performance, change faulty components, rebuild failed components, understand the intricacies of how each component works in its system, and learn the individual personality of each system in order to assist in ensuring its reliability. They learn how to make adjustments that are component specific or system reliant and not just what the nominal setting should be. There is a lot more to knowing how a system works and how to make it perform at its peak than understanding the theory.

DEGREE VS. APPRENTICESHIP

In comparison, a four-year engineering degree does not teach you any of this. You learn a lot about what things are and what they are supposed to do. You learn social science, English, math, psychology, and many other things, but you don't learn how to make anything work. Most of the time you are being taught by professors who lack practical experience and have nothing more than an academic knowledge. They offer theory but no hands-on experience about how to implement these theories and make things work. The belief that this is enough is compounded by the decision by senior leaders that this is enough to give you the tools to lead reliability efforts. Again, do not get me wrong, I have several degrees, but none of them in and of themselves grant me the ability to lead reliability. It is my journeyman level skills that do that. It was the shift in senior leaders' decision making that has led to the hiring of engineers to lead reliability that started the death spiral. At first it did not have a big impact because there was still a lot of bench strength in the trade ranks. This covered for the lack of trade skills possessed by these engineers and unqualified leaders. However, over the last 30 years this approach has eaten away at maintenance and reliability like a cancer, until we arrive at today, with the continued practice of hiring engineers causing the trade bench strength to leave us. This combination is killing reliability because there is no one left to develop the next generation of skilled tradespeople and if left unattended this will eventually lead to its death. The fix is for the few remaining reliability professionals to do something to turn this around. The first step is for our industry to understand that we have a problem. This is tough because it requires a large portion of today's leaders to admit that they are not qualified for their positions. The next step is to be willing to do something about it.

On a positive note, there are some who know they lack these abilities. The problem is that we turn to miracle drugs for a quick cure, in the form of consultants. This makes things worse, because the majority of consultants suffer from the same disease. They are engineers without the trade skills, spouting fancy sounding programs with quick fixes. They all use the terms "best practices" or "world class," but these are little more than empty promises rooted in no real substance. There is no substitute for hard work. Copying what consultants say they achieved somewhere else will make them rich and you worse off than if you did nothing. Their goal is to get paid, not your organization's success.

You do not gain a competitive advantage by copying others. True leaders *create* and do not need to copy. Each organization is its own entity, and you cannot copy your way to success. Reliability leadership needs to consist of people who can create on their feet using years of experience and education. The respect of the reliability personnel is necessary, and that comes from the shared experience of doing the jobs we are asking them to do. To lead the changes necessary in any organization, you must have a leader who has done the work and is not reiterating a theory. They have held each reliability position and can train and develop the personnel needed to drive success.

TOP MANAGEMENT TEAM

With an ever-increasing competitive business environment, corporations need to take advantage of every opportunity to maximize their resources. The composition of a *top management team* (TMT) is of critical importance in accomplishing this. A successful top management team is usually comprised of members who have higher educational levels and have risen through the ranks. The tenured team members are normally made up of members that have risen only through the operational ranks, or worse, transferred from another organization or industry and having even less qualifications. The most recent wave in senior leader promotion is to elevate attorneys or accountants who have no practical manufacturing experience. They focus solely on leading by policy or the bottom line. The effect of the TMT composition has a direct effect on corporate strategy and competitiveness.

With the increase in corporate competitiveness and the need to do more with less, Top Management Teams are looking for ways to eliminate waste from their organizations and maximize the use of their existing resources. Opening their eyes to an untapped resource will intrigue not only TMT but reliability professionals looking to advance within the reliability arena and within their organizations. Where there was once a ceiling for these professionals, there may now be an opportunity.

When I get asked if leaders are made or born, I answer made. In the multitude of leadership training classes I have attended, I have heard more times than not people answer this question with they are born. I always chuckle because if that was true, why are we all in this class? I have included four major objectives for the leadership discussions in this book. The first is to provide a deeper understanding of what makes up leadership development. The second is to challenge the notion that leaders are born. The third is to give an explanation of what elements comprise leadership efforts. The fourth is to develop full leadership potential and create an ability to turn those they lead into leaders.

What is Reliability?

COMMON CAUSES OF MACHINE FAILURE

Equipment fails for a lot of reasons and not all failures are the same. The term "failure" or "malfunction" means that the equipment has stopped functioning the way it was designed to. I call this unreliability of the equipment or component. An example of this would be if a pump is designed to pump 10 gallons per minute, but as it ages its output is reduced to 8 gallons per minute, this is a form of unreliability. Using the Pareto Principle, we know that 80% of the failure occurrences are caused by 20% of the causes. The same is true if 80% of all downtime is caused by 20% of the plant's equipment.

Machines do not die—we kill them! Though equipment fails for a lot of reasons, they all fall into one of three major categories: improper lubrication, contamination, and improper installation. This book is designed to help you understand the pitfalls associated with efforts to eliminate these problems and how to properly address them using a hands-on, practical, common sense approach.

UNDERSTANDING RELIABILITY EFFORTS

To achieve reliability, an organization needs to understand that there cannot be a company or production plan and a reliability plan. There should only be one plan, and the name is not important. I refer to all plant operational efforts as reliability. If you visit any organization that I have worked with or in, you will hear and see a lot of reliability focused approaches. Integrating production and reliability is the only path to success. This is easier said than done.

Across most industries, the words *maintenance* and *reliability* have become synonyms. This could not be further from the truth. If your organization believes that these words are interchangeable, you have a big uphill battle. The difference between maintenance and reliability is vast in effort and focus. To maintain, you strive to keep the status quo. You are not improving your processes because you believe you have achieved your desired approach. A maintenance effort is something the maintenance department is responsible for. Most organizations that have maintenance departments are operating in silos and look for maintenance to do all things involving physical labor. The mindset that is predominant throughout these companies is "we operate it, and you fix it."

THE FUTURE

Reliability is focused on the future. A reliability effort is all about continuous improvement and looking for proactive measures to intervene in all failure modes to enable your equipment to reach its designed life. I will go as far as renaming your maintenance department a reliability department. This is a cultural change and as we all know changing people is tougher than changing an approach. One big change is to eliminate the antiquated view that reliability work is a service to production. These efforts are equals and partners to production. Without either you do not produce your product. One key element to making this change successfully is having a true reliability professional with real-world experience in leading change. The individual needs to have a transformational leadership style to effectively transform an organization. A true reliability professional will be able to articulate how to get this done. Not with fancy sounding programs or "mother statements" but by being able to share hands-on experience doing it, not repeating something they read or heard. We will discuss these individual items throughout this text.

Reliability is an operation objective and not a maintenance effort or flavor of the month. If reliability is implemented correctly, it becomes the organization's way of life. This requires support from every level of management. Only with the support of all departments will any reliability effort succeed. As I have said, and it deserves repeating, an important step in achieving this is to break up the supplier-customer mentality and to work as a *team*; otherwise, any reliability effort will not produce results and will frustrate everyone involved.

Reliability is an organizational level effort and to achieve this the organization must address the three main root causes that result in unreliability: improper lubrication, contamination, and improper installation. Of these three, lubrication, or better said, improper lubrication is the most prominent.

A quick visual reference of the breakdown of unreliability can be found in Figure 1.1.

How does this breakdown work? Improper lubrication drives mechanical wear and corrosion, which leads to premature failure of your lubricated assets. If your organization does nothing else to improve reliability, you should address your lubrication. It offers the biggest bang for your buck and can be implemented relatively cost effectively because most organizations are already spending

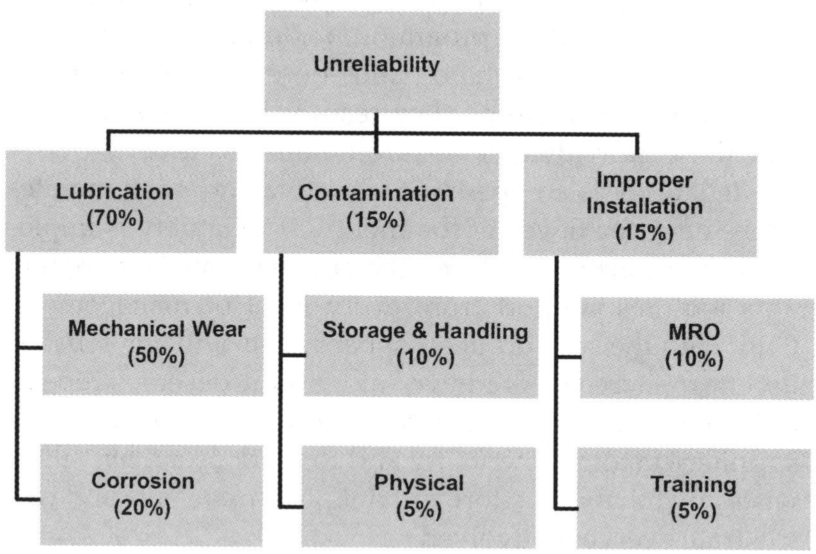

Figure 1.1 Unreliability Hierarchy.

money on lubrication and they just need to know how to do so effectively.

Contamination is the next group, which is broken into contamination of your lubricants and spare parts (specifically in their storage and handling), and the physical contamination of the equipment caused by poor housekeeping. Again, it is relatively cost effective to get your teams to do housekeeping and store your lubricants and spare parts correctly. However, it can be difficult to achieve if your leaders lack the necessary skills to lead their teams. Poor housekeeping is a direct representation of a lack of leadership, so addressing this problem needs to be focused on your leaders, not their teams. The third and final group is improper installation. This is made up of your maintenance, repairs, and operations (MRO), engineering, and your training. As I have repeated about the first two groups, this is a cost-effective way to be reliable. Inventory assurance is key to your MRO and the hub of your reliability. You are already spending money on the inventory, all you need to do is store it correctly. Your MRO is part of your contamination control program as well, so you can see that these three groups are all intertwined. Training, or more importantly, *development* of your teams is what will bring this all together. Without a properly developed team, none of these efforts will be sustainable. Improper installation starts with your equipment design. Those charged with the design of your equipment need to understand reliability and design it into your equipment first and foremost. Designing the equipment for reliability and maintainability will set your organization up for success.

Reliability is all about getting out in front of your equipment, understanding its failure modes and how to stop them from happening. A lot of elements cause equipment failure and because of this about 90% of failure happens randomly. You can try to prevent failure by changing your parts before they fail; this works 12-18% of the time. You can attempt to predict their failure but this is only 50% accurate. To achieve reliability, you need to be proactive and stop improper lubrication, control contamination, and eliminate improper installation. This will stop 90% of the random failure and unreliability. Now, to borrow from Stephen Covey, we must start with the end in mind, and we will start by discussing the biggest bang for your buck, improper lubrication.

CHAPTER 2

LUBRICATION

Lubrication is a must with respect to reliability, but what exactly is lubrication? Webster's Dictionary defines lubrication as the application of some oil or grease substance to diminish friction. Although this definition is valid, it does not encompass everything that proper lubrication can achieve. It is this simple definition that often explains why organizations place little or no importance on lubrication. They lack a thorough knowledge of reliability and how lubrication impacts it. I am giving organizations the benefit of the doubt here, because if they understand the importance of lubrication and do nothing to ensure its complete, proper application, they would be criminally liable. Think of your lubrication systems as the circulation system in your body. To stay alive, you need to do a lot of proactive reliability efforts.

The driving force behind the need for a lubrication program is the fact that 70% of all unreliability is caused by a lack of a lubrication program. Improper lubrication causes mechanical wear and premature equipment/component failure, or unreliability. Figure 2.1 illustrates the major causes of unreliability and the breakdown of components.

Figure 2.1 Improper Lubrication.

MECHANICAL WEAR

Mechanical wear of equipment components is common, and this is particularly true where improper lubrication is an issue. Particle and moisture contamination, along with the wrong or degraded lubricants, are the prevalent factors in creating rust on metal components. This increases oxidation rate, which leads to increased acid within the components.

Mechanical wear happens when machine surfaces rub against each other. Abrasive wear happens when particles enter the system, commonly through contaminated lubricants. These particles are usually dirt or wear materials, and they lead to three-body abrasion known as surface fatigue, which pits and scores the machine surfaces. This results in premature failure, also known as unreliability. Adhesive wear results when two surfaces come in direct contact and transfer material from one face to the other. This happens because lubricants cannot support the load, or in areas where the surfaces suffer from lubricant starvation or the use of the wrong lubricant.

Metal fatigue is another form of mechanical wear. An example would be when you work a wire back and forth to cut it without tools. The more you move a wire back and forth, the harder the metal works; the fatigue increases, ultimately resulting in a brittle point that snaps. This is the effect contaminants have on metal surfaces. Over time, this constant flexing will fatigue the metal, resulting in premature failure. The Institute of Mechanical Engineers has found that "every $1,000 invested in proper lubrication yields a savings of $4,000, which is a 3,900% return on investment." This is a great return on investment in anyone's book, and an example of how to justify a project and fund your reliability efforts.

CORROSION

Acid is a by-product of the reaction between additives in the oil and water. Lubricant contamination has many paths. The manufacturing process of the lubricants is the first-place contamination enters the system, but it can enter through seals, creating a caustic environment that results in wear. This can also happen if you do not understand the lubricant's additive package, because you can end up with corrosive damage. An example would be using an extreme pressure additive with yellow metal (copper, brass, etc.) because some of these additives are corrosive to them. This is another example where you need someone on your team who understands this problem and is qualified to review and eliminate it from happening.

Equipment loses its functionality in three ways, but it is the lubrication (life blood) of the parts that is the root cause of much of these failures. Keeping your plant equipment properly lubricated, controlling the contamination, and ensuring the proper installation can extend the equipment life and contribute to a more reliable operation. To proactively address the single most common cause of unreliability, you need to develop a lubrication program.

To start a lubrication program, you need to understand that there are several different substances that can be used to lubricate a surface. The most common are grease and oil. Grease is comprised of oil and a thickening agent. The thickening agent's job is to retain the desired consistency, while the job of the oil is the actual lubrication. The selection of the proper grease should be done with the utmost diligence to address all equipment needs. Many of the manufacturers sell inferior grease, and the thickening agent and oil separate. This leads to lubrication-based failures because once the oil separates, the grease loses its protective quality. I have seen organizations use extension hoses to make grease points easily accessible. This can be a machine killer depending on your grease. You need to know the amount of time the grease will sit in the line based on the frequency of application and amount. The wrong grease will allow the oil to leach out of the thickening agent and all you will be providing the lubricated asset is the thickening agent, devoid of oil. What would appear to be a great idea to an uninformed organization could be a self-inflicted wound.

OILS

Oils come in three common varieties: vegetable, mineral-based, or synthetic. There can also be a combination of these as well. The application dictates which oil, referred to as the base oil should be used. Synthetic oils are designed for extreme conditions while vegetable oil is used where there are environmental concerns. Your decision on which to use needs to go beyond application and should consider your overall program. Resources will play into your decision as the different bases require different proactive approaches. If your organization plan includes in-the-field filtration instead of interval-based replacement, you need to understand the effects on your base and its additives. Reliability leaders need to have experience with this or you will leave this type of decision to a supplier who may be unqualified and most certainly does not have any "skin in the game." Most organizations leave this important decision up to their supplier because they offer "free" advice. Is it truly "free" if your reliability suffers? Part of your justification for your program start-up is the real cost of "free" advice and its detriment to plant reliability. Suppliers have very little risk in providing free bad advice because if an organization is allowing them to give it, the organization lacks the ability to determine that it is bad.

ADDITIVES

In addition to determining the correct base oil, the selection of additives will enhance, add, or suppress properties within the base, and the lubricant's performance. The additive package is determined by the type of base oil and the application that it will be used in. An example would be engine oil with a dispersant added. The dispersant keeps insoluble matter conglomerated together so it can be captured by the filter during circulation. Additives can improve anti-wear and extreme pressure performance. It can enhance seal performance and reduce or eliminate start-up wear. Allowing the supplier to dictate such an important decision should never happen. Again, this seemingly "free" advice is costing you thousands or even tens of thousands of dollars in unreliability caused by premature failure of your lubricated assets. Educating yourself on machinery lubrication is vital to the selection of the proper lubrications, as is selecting a lubrication partner who has the same goals you have—plant reliability, not their profits.

In an application that experiences extreme temperature, from hot to cold, a viscosity index (VI) additive would be used. The molecular construction of these additives has long, organic molecules that stay together in cold conditions, but they unravel in hot environments. This allows the oil to change its viscosity and flow better when it is cold while maintaining its high-temperature properties. The main concern most organizations should have with additives is that they can be depleted, requiring the oil to be changed to restore them to required levels. This was the "old school," or supplier-preferred approach. In reality, you can re-infuse additives without a complete oil change.

Working with a lubricant partner, you can determine what portion of transfusion is required to return the additives to desired levels. The combination of in-the-field filtration, base oil, and infusion is another cost savings. Most supplier representatives are not qualified to assist you in this decision so they will focus their efforts on sales and not your performance or cost savings. This is another aspect of reliability where you can save money to fund other reliability efforts.

Friction

The key objective of lubrication is to reduce friction, but it has a lot of other benefits. The lubricating film helps to prevent corrosion by shielding the metal surfaces from water and other substances. It also has a role in controlling contamination within the system. It serves as a conduit to move the contaminants to the filter for removal. Another aspect of lubrication is that it aids in controlling the temperature by absorbing heat and transferring it to where it can be dissipated. Accordingly, the selection of your lubricants needs to align with your reliability strategy and is key to its success. Selecting your supplier cannot be left to purchasing agents and supply chain managers. They are not qualified to make this decision. Selling the importance of lubrication is the job of Reliability Managers, and they can only do so if they have the experience and education to communicate this at all levels of the organization. Do not entrust this solely to your supplier; educate yourself in machinery lubrication so you can instruct your suppliers and not just hope they get it right.

Lubrication has three different types: boundary, mixed,

and full film. Each of these is different, and they rely on both the lubricant and the additives to protect against wear.

There are two forms of full-film lubrication: *hydrodynamic* and *electrohydrodynamic*. Hydrodynamic is when two surfaces, relative to each other, in a sliding motion, are completely (fully) separated by a fluid film. Electrohydrodynamic differs from hydrodynamic in that the surfaces, relative to each other, are in a rolling motion. An additional difference between the two is that while electrohydrodynamic has a much thinner film layer, the pressure on the film is much greater. This is called electrohydrodynamic because the film elastically deforms the rolling surface to properly lubricate it. If you want to test your supplier, ask a basic question about this and see if they can answer it.

Asperities

No matter how polished or smooth the surface appears, there are always irregularities present. At a microscopic level, they jut out from the surface like peaks and valleys. The peaks and valleys are called asperities. It is the peaks of these asperities that carry the load. To achieve a full film condition, the lubricating film must be thicker than the asperities. This type of lubrication is the most desired and protects the surfaces most effectively. By filling in the peaks and valleys, you can spread the load across the entire surface area, reducing friction.

In applications where there are shock loading and frequent starts and stops, boundary lubrication is preferred. Additives like *extreme pressure* (EP) or *anti-wear* (AW) protect surfaces where a full film cannot be accomplished because of load, speed, or other factors. These additives form sacrificial layers of protection by clinging to the surfaces, preventing wear. When surfaces are contacting in such a way that only an EP or AW layer will protect them, boundary lubrication must be used. However, this is not the ideal situation because of high heat, friction, and other undesirable effects.

In applications where you need both boundary and hydrodynamic lubrication, you can employ mixed lubrication. In this application, most of the surfaces are separated by a lubricating layer but the asperities will still come in contact. Because of this, you will need to employ additives.

Only with an understanding of the process will it be easy to determine which lubricants to use. The determination is between separating the surfaces or protecting them from friction, heat, wear, and other issues. Using oils, grease, gas, or other fluids will accomplish this. Next time you change the oil in your car, understand that there is a lot more going on than meets the eye.

BUSINESS CASE

When making the decision to undertake any project, you must understand what constraints you have. The constraint can be displayed as a triangle. A graphical representation of this decision making is shown in Figure 2.2. One leg of the triangle is *costs*, the next is *time*, and the final one, *quality*. The middle of the triangle is traditionally risk. I like to look at these projects with an eye on the opportunity and not the risk. All projects have associated risk and opportunity. Your job is to look at the risk and mitigate it to the best extent possible, or to look at the opportunity and maximize it.

My recommendation to all organizations is to never compromise on quality. This is always the approach I take. If I cannot do it with the quality I set, then I will not undertake the project. For me, the decision on how to approach

Figure 2.2 Project Decision Making.

the project comes down to cost and time. If cost is the determining factor in your organization, then it will take more time. The reverse is also true: if you want results in a quicker time, then the costs will increase. The decision on costs and time needs to be made by looking at the opportunity. Organizations that keep their eye on the opportunity will gain a competitive advantage over their competitors.

When determining how much the potential savings will be when you undertake a lubrication project, you can use detailed information from your *computerized maintenance management system* (CMMS) or other cost tracking system or industry standards. Most organizations lack the detailed information to calculate the cost of not having a strong lubrication program, so I normally use percentages I have gathered throughout my experience. If your organization does not gather this information, you will need to determine potential savings.

ROI Calculation

Since most organizations today use the terms maintenance and maintenance budget, I will use those words throughout the example calculation to ensure ease of understanding. You need to start with the total cost of doing maintenance in the facility. This is accomplished by taking the annual maintenance budget and adding the cost of downtime. Most organizations have a cost associated with every hour of downtime, and you just need to simply do the math. This includes all unscheduled downtime costs (expedite fees, overtime, contractor overtime, etc.) as well as de-rated production (production overtime that is subsequently scheduled, loss of customers, missed delivery, etc.). Now that you have the total costs for maintenance you can begin narrowing it down to the opportunity provided by doing lubrication correctly. First, you take 50% of the total maintenance costs to determine the cost of repairs. This is what it costs to do repairs for all systems. If you track this figure you can use the exact number. If you are like most and do not know what these costs are you can use a percentage; I use 50% based on my experience. This is based on averages across all industries. It will also be driven by your organization's maintenance strategy. From this, you take 50% of the amount to identify the repairs associated with mechanical wear of lubricated components. This eliminates electrical repairs and any other non-mechanical repairs of lubricated assets. Again, if your organization tracks downtime to this level of detail, use your actual percentage, but if you are the norm and do not track these costs, I recommend using 50% based on my experience. Then you take 50% of that amount to determine the mechanical wear due to poor lubrication. These are the repairs needed as a result of using the wrong lubricants, too much or too little lubricants, ineffective contamination control, ineffective sampling program, etc. Like before, if you track these percentages, use your numbers. This 50% is the average I have used based on my experience. Now you need to take 70% of this amount, which identifies the lubrication-related problems that can be avoided by a good program. This one is simple. If you implement a lubrication program that meets or exceeds industry standards, you will save this percentage of your current costs. This could be a lot higher or lower depending on your current practices. Now you add 5% of the annual maintenance budget associated with the costs of lube inspections, analysis, and other non-repair activities. This is the average percentage organizations spend, so insert this percentage if you believe it represents your organization. Most of the organizations I have worked for spend this percentage on their lubrication activities. If you have a program of some kind this will be the money you are spending. If you track these costs use your numbers. If you do not have a lubrication program of any type, you would take a smaller percentage here. In my experience, if you are only buying lubrications and pouring them into your equipment as your program, you are probably spending 3%. The final percentage is to take 30% of the cost of lubrication activities that happen as a result of lube inspections, analysis, and other things that are wasteful. These costs are associated with unnecessary oil changes and other unnecessary activities related to lubrication. If you use these universal truths about lubrication, you can quickly have the savings opportunity by implementing a good program. An example of this calculation using a one million dollar maintenance budget can be found in Figure 2.3.

An interactive software containing this calculation is available at *www.deathofreliability.com*. This software can be modified to determine return on investment and justify any project.

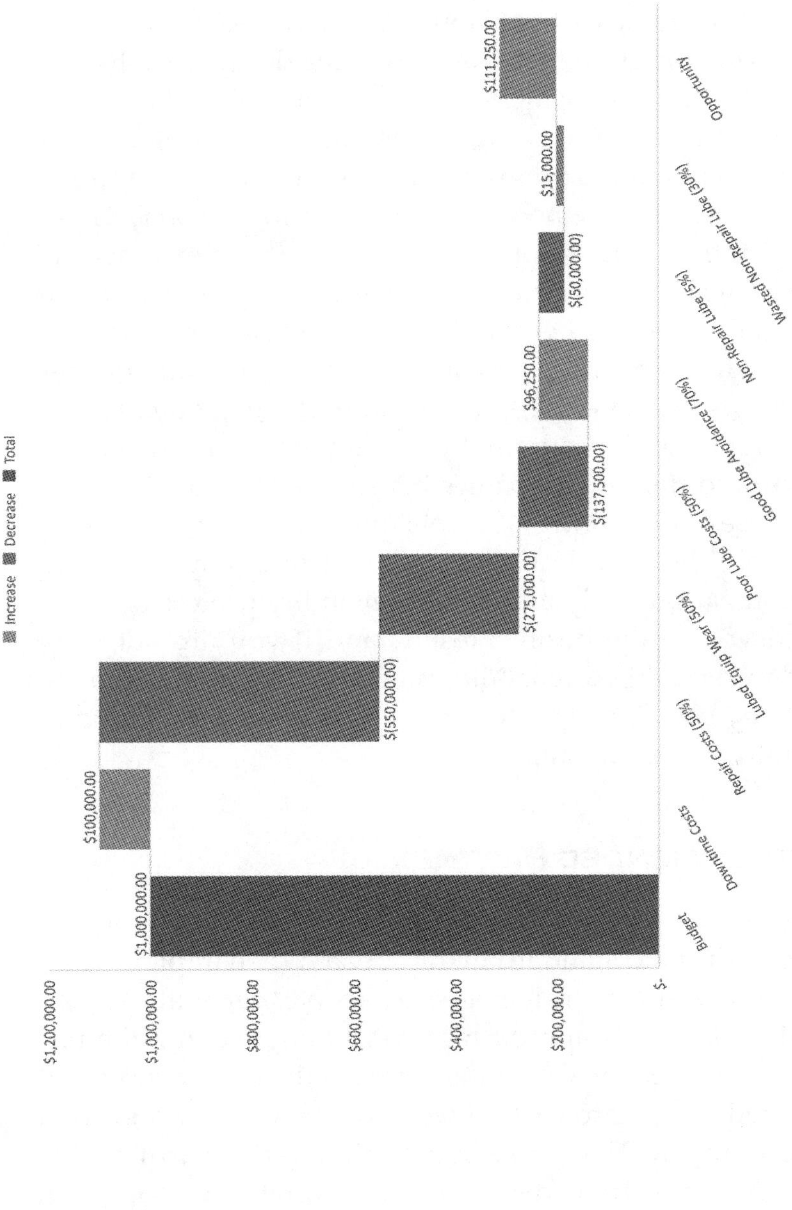

Figure 2.3 Justification Calculation.

PARTNERSHIP

There are some distinct steps necessary to achieving a lubrication program that will deliver the desired results. You need to build a business case by determining the costs of benchmarking your site, gathering data, and designing and implementing the program. The first step you need to take is to benchmark where your organization is today. If you have the internal resources to conduct this benchmarking and time is not an issue, you can use this method to accomplish this. I would argue that the best way to accomplish this is to collaborate with an outside subject-matter expert who has the expertise to help you pull together all aspects of a lubrication program. In my lifetime, I have dealt with several different companies boasting of this ability, and I have seen only one that meets my expectations. Just to be completely clear, I am not affiliated with any consulting organization. Only a qualified reliability professional will lead you to the desired results and if you are not being led by a qualified reliability professional, you need to hire that person first. The comments that follow are based on results, not friendship.

RECOMMENDED PARTNER

Based solely on my proven success, The Noria Corporation is the gold standard in this arena. When implementing reliability efforts, it is necessary to partner with experts. This is a scalpel approach to consultant use, not the normal meat cleaver. Consultant use needs to be focused and limited in order to combat weakness in your organization. I recommend The Noria Corporation as this partner. They can accomplish in weeks what would take years if I were to dedicate internal resources or entrust this important task to any other consultant. My father gave me advice a long time ago that I have lived by for all my years. He told me that I do not need to be the smartest man in the room, but to succeed I need to surround myself with them. In my playbook on eliminating unreliability, The Noria Corporation is my go-to organization in this area and absolutely has the smartest people in the room. Since lubrication is the most impactful of the three things that cause unreliability, your selection of who to collaborate with in this endeavor is the most important decision you will make. I rarely use the term never but when it comes to selecting an organization to collaborate with and ensuring that my equipment has the best blood system possible, I never make the decision based on costs or whim. You would never select a cardiologist for your child based on either of these, so never do it when it comes to the life-blood of your equipment. To determine the costs of developing a comprehensive lubrication program, you must work with your partner. The process I have used, and trust, is listed below. The Noria Corporation can provide you these cost quotations and they will not be the cheapest or the most expensive, but I guarantee you they will be the most value added. Having done this for many years, I can tell you that they are the best value for your money.

To calculate a return on investment, you will need to know the costs of the phases required to implement the lube program. You need to identify the costs associated with a benchmarking assessment, data gathering and design, designing a storage and handling facility, and training your team in proper lubrication. Then you need to develop your program and team into a proactive juggernaut.

Every organization has its own method of determining the viability of a project. Now all you need to do is put this into the calculator used by your organization. Most organizations I have worked with use the information listed in the following section in one form or another. I provide this for reference, and it is up to you to work through your organization's process. There is an interactive program that can be used to calculate your return on investment located at *www.deathofreliability.com*

DATA GATHERING

It is essential to have someone verify every lubricated asset on your site, and this is a time-consuming task. Without the expertise to accomplish this quickly, you will expend many more hours in this effort, and it will require you to continuously retrace your steps because there will be missing information that will require repeated visits to the assets. Consequently, the most cost effective, time saving, and quality effort you can put forth to accomplish this stage is to collaborate with an expert. Don't get me wrong; I have my internal team work alongside the partner because these efforts will go through many iterations, and hands-on development of my team is key to continuous improvement. Also, paying for your partner to return with each iteration will eat away at your savings. For scoping purposes, I would think the data gathering and design phase of this project will look like this.

There will be both on-site and off-site work associated with this project. The on-site work will consist of a kick-off meeting between the company sponsors, the partner organization, and employees with knowledge of the plant and its existing lubrication program. The next step is to swap and review the available data related to the program with all parties. Survey the current lubrication storage and handling system, which will aid in the design of the future storage and handling facility and program. Finally, there will need to be a detailed, extensive, hands-on data gathering phase. Every effort needs to be made to minimize downtime for this, and there should be no unscheduled downtime.

The off-site work will consist of processing the data collected during the asset walk to make decisions about each lube point, contamination control, and lubricant analysis. This step is followed by confirmation of, and needed changes to, the recommended lubrication types and grades. Next comes the development of hardware adaptation proposals and practices for all equipment where it is necessary to attain ideal maintainability. This is followed by recommendations and practices for enhancements to component contamination control ideal practices. There needs to be a development of tailored reference practices for all lubrication proactive reliability and *condition-based monitoring* (CBM) activities, to include job-specific elements, like grease or oil quantity, filter efficiency, etc. Each task and recommended interval will be assigned based on the lubrication, *Original Equipment Manufacturer* (OEM) and supplier preferred practices. During this off-site work, there will be a grouping of machinery data in practical location hierarchies to assist with the establishment of lubrication routes. The next step is the development of oil analysis test schedules and sampling intervals for all oil-lubricated equipment as part of the oil analysis program. This is followed by an on-site review by your team to look for conflicts or equipment-specific nuances that your partner may not be aware of. Once this normalization is done, you will have a foundation for a proactive approach that will begin to immediately save you money and impact your reliability. The hard work begins with the implementation of all of these findings. I have found that this is where most organizations stop, because they lack the qualified reliability leader to get them over this hump. As I have mentioned before, there is a big difference between knowing and doing, and without a qualified leader you will be left with knowing what you should do but not how to do it.

LUBRICATION PROVIDER

Organizations that make the mistake of allowing their lubrication provider to do any of this work without verification by an organization like The Noria Corporation does so at their peril. Every case that I have seen where the provider was also in charge of the design ended with substandard results. Providers will offer free studies and other services, which are enticing to organizations who do not know better and are budget driven, but if this is the path you take you should save your money. Few providers have qualified representatives to accomplish what is needed. Without a qualified reliability leader in your organization who knows the difference, you will not achieve the desired results. However, having said this, I do have a preferred provider of lubricants as well. The comments in the following area are again based on results and not friendship or any other personal relationship. Qualified reliability professionals align themselves with the best in the industry and their decisions are based solely on performance. Based on my experience, Schaeffer's Oil is the provider partner I always turn to.

Recommended Partner

In my experience, I have had the opportunity to test different lubrication companies against each other on like equipment in a like environment. On every occasion, one provider stood above all others. This provider even helped the United States, free of charge, win the first Gulf War. When our tanks and aircraft were grounded by sand contamination, this provider stepped in and provided a lubricant that eliminated this contamination and returned our equipment to service and victory.

Tested against other providers, there was a clear winner in all aspects. The winning provider was Schaeffer's Oil in St. Louis, Missouri. Schaeffer's Oil is the oldest petroleum provider in the U.S. The company has the largest number of certified lubrication specialists in the industry and has been family owned and operated since its inception. There are several reasons they stand far above the rest. Schaeffer's offers new concepts in lubrication and reducing operating costs. They accomplish this by reducing total reliability costs. The focus needs to be on "cost to use" not "cost to buy." In my practical experience, using Schaeffer's lubricants has reduced my organization's lubricant related costs by 30%. This is accomplished because Schaeffer's uses the latest technology in base stocks and additives. They combine the finest quality synthetic base stock and 100% pure paraffin base oils, resulting in a product that withstands higher heat range and operates cooler. This is a huge cost savings and all by itself justifies your switch to Schaeffer's. A competitive advantage that

Schaeffer's offers, which their competitors cannot, is their proprietary additives containing an adhesive-cohesive that allows the lubricant to steadfastly cling to the gears, seals and bearings. This guarantees that the Schaeffer gear oils retain a protecting film that "stays put" on the metal surface of the gears and bearings in spite of how thoroughly it is wiped away. As we have discussed, mechanical wear is a significant issue in unreliability, with 77% of all mechanical wear happening at start-up. It is caused during the time when the mechanical components start to turn and lubricant has not reached these components. Schaeffer's additives cause their lubrication to reach these areas immediately upon startup because of their capillary effect. The lubricants literally climb all mechanical parts, providing lubrication in advance of the system's ability to lubricate these components. Schaeffer's lubricants also contain the additive Penetro, which forms a protective layer on all metal surfaces that in turn reduces the friction and leaves a protective layer to protect the metal during startup. This specifically targets the 77% of mechanical wear and ultimate unreliability of your equipment.

THE GOAL

The goal of your lubrication program is to keep your equipment clean, cool, and dry. For every 10 degrees Celsius or 18 degrees Fahrenheit you increase or decrease the component life by 50%. If you have the right program storing, handling, and delivering the best lubricants to your equipment, you can do more with less just by changing your lubricant suppliers. Based on my experience, Schaeffer's Oil is the only lubricant I will recommend or use.

As the oldest petroleum provider in the U.S., Schaeffer's offers support that is unparalleled by their competitors. They are owned by the family of founder Nicholas Schaeffer, and their customer service focuses on providing you the best solution, even if it is not Schaeffer's. I have dealt with a lot of petroleum providers, but only Schaeffer's has ever cared more about my lubrication issues then their profits. They also offer a hands-on assessment of every lubricated asset and a review of all associated original equipment manufacturer (OEM) manuals to ensure that their recommendations are in the best interest of your equipment. When you pair this with The Noria Corporation's offering, you end up with a completely accurate, comprehensive lubrication program that will eliminate unreliability due to improper lubrication.

Lubrication is the single largest opportunity in reliability. When developing a lubrication program, you need to partner with the right folks to achieve the desired results. As with all projects, you need to build the business case and once you have identified the funds, through savings, necessary to execute the program you can follow the plan above to guaranteed results. Helping your organization understand the importance of lubrication may be a lot of work. As I have mentioned, initiatives like lean downplay the importance of lubrication, even delegating it to unskilled operators. As we have discussed, lubrication requires a skilled professional who understands not only how to do it, but why. If you allow your organization to turn lubrication over to unskilled labor, you will suffer the fate so many before you have suffered.

CHAPTER 3

CONTAMINATION

Proactive reliability is all about keeping your equipment clean, cool, and dry. The importance of keeping things clean cannot be overemphasized. Contamination will attempt to enter your operation at every opportunity it is given. To combat this, you need to be out in front of where it can enter. Most contamination will come into your operation through the entrances and exits, but your operation itself may also be a source. After years of working to eliminate contamination, I have pulled together the following areas that will help you along this process.

STORES

The start of your contamination prevention program should be your storeroom. The manner in which you receive and store your repair parts will set the stage for your plant's reliability. This contamination control starts with your vendors. If you do not have a specification for how you will receive your spares, you are missing a key aspect of obtaining reliability.

Your specifications must start with your cleanliness standards. The parts or materials you are receiving will drive these standards. You should visit the vendor and understand how the parts are being stored before their delivery to your plant. At no time should you allow a part to enter your facility if you do not know what conditions it has been subjected to. If you miss this opportunity to improve reliability in your operation, you are allowing your vendors to set your standards. I am always disheartened to hear maintenance and reliability leaders tell me that they cannot control what happens to the parts before they enter their facility. This cannot be further from the truth.

Once the part has entered your facility, you need to make sure it is cared for as well as all the other parts of your operation. All too often, organizations pay no attention to their spares until they install them. This is too late. The parts in your storeroom must be treated and maintained as if they are already installed in your equipment. The storage of gearboxes, motors, belts, bearings, cylinders, etc. will dictate your ability to deliver on reliability. Depending on your organization's manpower and stores facilities, you may want to review and decide where and how all your spare parts are stored. It is not necessary to have one or more of everything in your storeroom. I have used a very specific process flow diagram to determine what to store that has served me very well. Figure 3.1 is the flow diagram. The decision on what to store is one that requires a lot of discussion and should be devoid of emotion. There are significant costs associated with storage of parts. The average is 20% of the cost of inventory, so if your stores value is one million dollars, it costs you $1.2 million

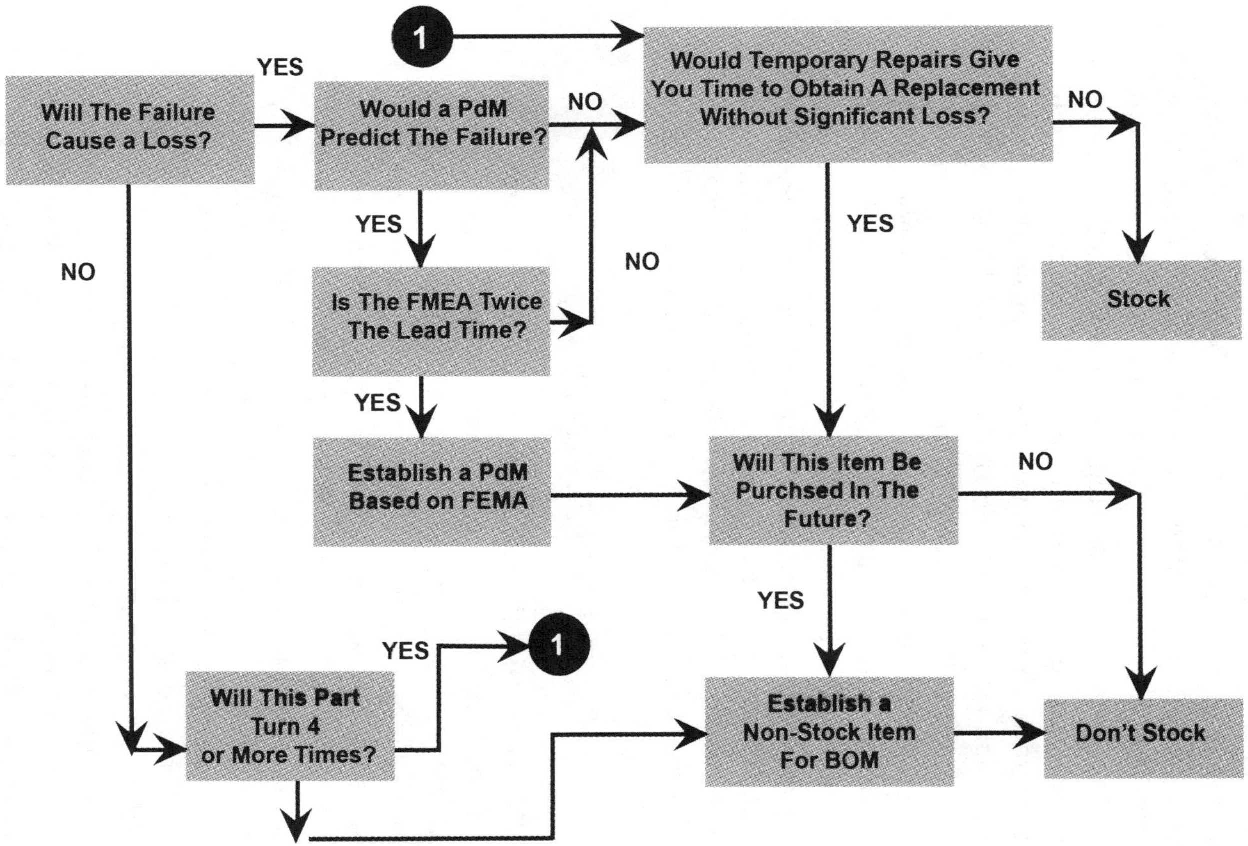

Figure 3.1 Stock or Do Not Stock.

dollars to keep those stores. If you stock the right stores and eliminate all duplicates and spoiled stores, you save money and obtain another funding source for reliability efforts. This affects your organization's free cash flow, and honestly, if you are not storing the parts correctly, it costs you even more money when you install them and they fail prematurely. The cost of downtime will prove that if you do not maintain your stored parts in a reliable state, you are better off letting your vendors store the parts correctly at their site and delivering them to you when needed. This may sound extreme, but if you add up all the time you spend storing parts incorrectly and compare it to the reliability of having them stored correctly, you will see that you make more product and spend less on replacing new parts gone bad. Of course, no one would do this, but it emphasizes the need to store your spares in a reliable state.

LUBRICANTS

Another contamination control point is the intake and storage of your lubricants. Figure 3.2 is an actual example of an organization's lubrication storage prior to my implementation of a lubrication and contamination control program. Most people believe that new oil is clean, but this cannot be further from the truth. Most new oil is 128% more contaminated than the maximum contamination level required by your equipment. The intake and storage of your lubricants needs to be a focus of your contamination control program. As you can see, the three major contributors to unreliability all intertwine. Improper lubrication, contamination, and improper installation are very interconnected, and our discussion will bounce between all three contributors but focus on the subtitled area. Because of this interconnection, it may be necessary to move between chapters when im-

Figure 3.2 Lubrication Program Before.

plementing my approaches to elimination of these causes of unreliability.

CLEANING AND INSPECTING

When we are discussing the cleaning stage, it can be compared to the steps in *5S* (sort, set in order, shine, standardize, sustain). Both have very similar purposes. The goal is to remove all grime and dirt from the equipment, not to make the equipment pretty but to reveal the issues hidden beneath the grime. To accomplish this, you will need to take the equipment out of production, remove all guards and covers to complete a thorough inspection, and drain all fluids. The goal is to return the equipment to "base condition." This is the best possible condition for the equipment's age. It is not intended to return it to new condition. It would require too much money to accomplish this and is normally not cost effective. Understanding "base condition" makes it possible to highlight the wear and damage to the equipment and develop the necessary repairs. Kaizen (improvement) is a lean term used to identify problems and either immediately correct them or schedule them for correction.

The accomplishment of the clean and inspect phase should be done by operators and reliability crews who are personally responsible for the equipment. This will increase their understanding of the equipment, such as where problems occur and grime accumulates.

REMOVE CAUSES OF CONTAMINATION AND IMPROVE ACCESS

During the clean and inspect phase, we need to identify where the dirt is coming from; is it generated internally or externally? The goal is to eliminate or minimize the source of the contamination that creates the need to clean the equipment.

Another portion of this cleaning and inspection is to identify hard-to-reach places or those that are unsafe. It is necessary to engineer out these hard-to-reach places in order to improve the next cleaning and inspection, and to accomplish this as quickly as possible. Some examples of this would be to install Plexiglas, or other see-through materials to facilitate inspections in the future.

EQUIPMENT CLEANING AND LUBRICATION STANDARDS

This first step identifies what work can be accomplished by operators to prevent further deterioration of the equipment. During the kaizen (improvement) event, the team defines what the operators will clean and inspect and how they will do it. Again, I stress that only employees who have been properly trained in lubrication should be allowed to lubricate your equipment. It will define how often it must be done to keep the equipment in optimum (base) condition. Figure 3.3 is an actual example of an organizations's lubrication storage *after* my implementation of a lubrication and contamination control program.

TRAIN/DEVELOP FOR GENERAL INSPECTIONS

Utilize the subject matter experts (reliability technicians and team leaders) to train the operators in their tasks and transfer the knowledge necessary to keep the equipment at base condition, teaching the operators to problem solve using tools like *"The 5 Whys."* The "5 Whys" is a root cause analysis tool used by many organizations to determine the root cause of their failures. Once the training is complete, the operators should inspect the equipment using their training to identify problems, just like during the kaizen event. Training the operators is not the end but the beginning. Developing the operator is how you make these efforts sustainable.

Controlling contamination is a key part of reliability. Contamination control is a term that encompasses all activities aimed at eliminating equipment contamination. This can be accomplished by reducing or eradicating contaminants through housekeeping efforts, resulting in more effective production. The most common types of contamination are: dust particles, wall and floor materials, packaging and crate particles, fibers, shavings from moving parts, other poorly maintained surfaces, liquids not being

Figure 3.3 Lubrication Program After.

cleaned up and equipment left damp to rust, and spills not being mopped up. These and many other damaging contaminants can infiltrate critical equipment areas in many ways. The goal here is to find and eliminate them all.

Contamination presents a significant risk to production activities. The proliferation of contamination will lead to product damage, recalls, and yield reduction, which is all detrimental to the business. Businesses are not protecting themselves from the effects of contamination, and their products are suffering from these unsafe processes. Eighty percent of all contamination enters the process through entrances and exits.

Therefore, one of the first logical steps to proactive reliability is the implementation of a contamination control program for lubricants and spare parts aimed at their entrance into the building.

THE STEPS TO CONTAMINATION CONTROL

Heat, moisture, air, and particles rob equipment of life, but with rigid contamination control practices, fluids can last indefinitely. This, in turn, prolongs the life of the equipment's components and keeps the machine running at the highest level of efficiency.

Additionally, the costs to start a proactive contamination control program are covered by reliability cost savings. A contamination control program can be employed in three steps.

1. Start by setting target fluid cleanliness levels for all equipment.
2. Select and install filtration equipment (or upgrade existing filter rating) and contamination prevention techniques to achieve the cleanliness levels.
3. Monitor cleanliness at pre-determined intervals to achieve the cleanliness levels.

CONTAMINANT MONITORING: THE CORNERSTONE OF CONTAMINATION CONTROL

For the same reason you do not drive a car without a working fuel gauge, you shouldn't attempt proactive reliability without a monitoring program. Monitoring will give you the data you need to ensure that your machinery is operating within contamination standards.

Proactive reliability is different than the other three maintenance approaches because it addresses much more systemic elements of a reliability program, rather than inspecting the machine itself. This approach is much more diligent and looks to manage the difficulties that can contribute to equipment wear and tear as opposed to the failure itself.

One of the key parts of any proactive reliability approach is controlling contamination. I have outlined in this chapter the accelerated deterioration you will experience because of contamination. In the previous chapter on lubrication, contamination was a significant factor in improper lubrication, so the interconnection between the three causes of unreliability is clear. Stop contamination and you will increase reliability. Stopping contamination does not cost a lot of money, but it does take a lot of hard work. I know I keep mentioning those dirty words—*hard work*— but doing the right thing is rarely easy.

Machinery Lubrication Magazine reported that a proactive reliability strategy looks to examine and fix the root causes of failures, like contamination. Though this needs some due diligence, it is a less expensive approach than the other approaches.

Improper Installation

Improper installation is the third of three causes of unreliability. To remember how to combat this portion of unreliability, I think of it as having the right person with the right part at the right time. Improper installation is made up of several parts; start your elimination of unreliability before the systems enter your building. You must design for reliability.

DESIGN FOR RELIABILITY

Designing for reliability is the most recent term for engineering the equipment for reliability. Ensuring that there are no hard-to-reach points that require the equipment to be shut down to service will increase maintainability and as such will achieve reliability right out of the gate. It is during this phase that the importance of an experienced reliability professional is clear. Without the experience to know what failure modes or other issues will arise with the equipment once it is installed in your plant, you will spend a lot of money and lose a lot of productivity after the fact when these failure modes and issues arise during production. Organizations fail to recognize the importance of this phase in ensuring reliability and entrust it to young engineers or senior engineers without the experience to eliminate these issues before they arise. This can be made even worse by handing it over to suppliers.

Providing the equipment manufacturers with your organization's specifications will go a long way towards setting up your facility for success. All too often organizations are more focused on the short-term cost of the equipment, and they pat themselves on the back for getting it cheaply. This is another failed approach that is prevalent in all industries today. Again, this is the result of leaders who lack the basic experience to run a plant effectively. They perpetuate the failed approaches of those who came before them because they are not qualified for the positions they are entrusted with. It will cost you less and you will get more reliability if you fix this problem before it becomes a reliability issue in your facility.

One of the best arguments for establishing these standards and specifications is standardization throughout your facility. Requiring the manufacturers to build to your specifications and not theirs will pay for itself hundreds of times over, absorbing any increased costs associated with the requirement. Equipment manufacturers use the cheapest materials they can get away with in their construction to increase their profit margins. If you do not demand quality, you will not receive it. By standardizing, you will reduce your development costs and materials costs. If you have little variation in your plant equipment, you need to stock fewer parts and your operators and technicians need to have knowledge of fewer systems. All of these things result in lowering the costs of operating and maintaining the equipment. Demanding that all the equipment's service points be accessible while it is operating will require fewer non-productive hours to proactively maintain the equipment. If your return on investment is calculated by the equipment's productivity, then maximizing the hours it can run will give you the best results. Sadly, the ability to design for reliability is not something taught in engineering schools. All of the thousands of engineers I have worked with miss this key aspect of designing and deciding on what equipment comes into your plant. This supports my declaration that the lack of senior leaders with a real reliability background is one of the driving forces behind the organization's inability to truly have reliable production.

The final stage in design for reliability is to conduct a *factory acceptance test* (FAT). This refers to the functional test that is performed by the vendors once they complete the assembly of the equipment. This test will verify that the equipment meets the specifications and functionality agreed upon in the purchase agreement. Before the equipment reaches the plant, it is imperative that an organization put the equipment through its paces and recreate all failure modes during this phase to work the bugs out in the factory and not in the plant. I will bet most of you reading this are saying to yourselves *"that is what we always do."* That is, you and your organization always seem to struggle for months or longer working the bugs out when you should be producing a product. The rush to get the equipment is shortsighted and something that poor leaders decide to do every day. Try as they will to justify this, it is factually a bad decision in all cases. There is no honest justification to rush the equipment into operation, and it will lead to continuous issues with the equipment for the entire life cycle. If you are telling yourself that I am wrong and your organization has done this successfully, you are lying to yourself.

SPARE PARTS MANAGEMENT

The next part of the improper installation is the management of the organization's spare parts storeroom. If the storeroom is run improperly (poor inventory, stock outs, etc.), the rest of the reliability operation does not stand a chance. The level of service in an organization's storeroom is a solid indication of the effectiveness of its reliability efforts. Having the right parts is key to the elimination of improper installation. Without the right parts at the right time, reliability is not possible.

What is the impact of not having the right parts? One thing is that the delay caused by trying to expedite the right parts to the plant increases downtime and costs. If this delay is beyond acceptable time restraints, then it becomes necessary to use an alternative part to make it work. This increases downtime and costs as well. Also, it requires additional repairs to remove the alternative parts and install the correct parts once they arrive. Now you incur more downtime, costs, and wasted reliability resources that distract the team that could be working proactively to address other issues. Of course, downtime is a loss of production and finished product; this impacts customers and their confidence in the organization.

The difference between this and other books is that now that I have offered up the causes of unreliability, I want to work on a solution. The sections that follow provide methods and practices for achieving storeroom excellence, which in turn enable excellence in the entire reliability operation.

The Storeroom as Partner

When you conduct a value stream map of the technicians and materials in most reliability operations, there is a common theme. You will find a central hub with spokes radiating from it like a wheel. The hub is the storeroom, while the spokes are the paths to the equipment in the plant where reliability is being performed.

The main function of the storeroom is to provide parts accurately when needed. It plays much more than a pivotal role; it is the *hub* (center) of successful reliability operations. Much in the way that most organizations view maintenance as a service, the storeroom is often seen as a service provider, with reliability (and by extension operations) its customer. This cannot be further from the truth. The main ingredient for success from the storeroom is to continuously monitor the needs of reliability and to adapt and partner with it. To manage this relationship, the storeroom needs to be deliberate in its communications. Reliability must know what assistance it will receive and in what fashion to count on the storeroom.

From the reliability standpoint, there is an expectation of parts availability. Most of the time this expectation can be met. However, when it comes to certain parts, this expectation is not realistic due to costs. Reliability management and the storeroom need to communicate to operations in these situations to effectively manage their expectations. Setting storeroom expectations after analyzing failure modes and their impact on downtime is the best method to use to ensure reliability. The best time to decide what parts are to be stocked is when the equipment is new and before it is placed into service. The original equipment manufacturer, the part suppliers, and the reliability leaders should work together to determine necessary spares. Once the equipment has generated historical data usage, reliability can adjust inventory levels and increase or decrease stock on hand. In larger organizations, a parts inventory planner can enhance the determination of stocking levels, re-order points, and replenishment trigger levels. The process that follows is a common storeroom evaluation that can be employed to ensure your organization is able to meet your reliability needs.

Scope of Work

Having an inventory that reflects and supports the current operation is one of the most important steps you can take to improve cost, work efficiency, and reliability. Having the right parts, in the right place, at the right time, and in the right condition is the objective of this project and scope of work.

Scope of Services

This project will combine resources from the organization and the successful partner to focus on the goal of cleansing the organization's inventory. The scope of work includes providing overall project coordination in fulfillment of the deliverables listed in the following section; organizational and administrative assistance to the Project Team and Work Groups; selection and coordination of Field Staff, and report writing.

Deliverables

The standard seven deliverables I use to accomplish this start with:

- Deliverable 1—Create a data format. I usually give the partner vendor two weeks from contract implementation. It is important that the format matches the computerized maintenance management system needs for bulk upload.
- Deliverable 2—Cleanse the current inventory data. Use the current inventory list provided by the organization to the successful partner to complete the following:
    - *Identify duplications, cross reference original equipment manufacturer (OEM) parts to the actual vendor for duplication identification and competitive pricing, work with key vendors to gather data in the correct format, and cleanse the data.*
- Deliverable 3—Cleanse historical purchase data. Use the historical purchase list provided by the organization to the successful vendor to complete the following:
    - *Compare current inventory to historical purchases, does inventory mirror historical purchases, and develop a variance list.*

- Deliverable 4—Cleanse historical emergency purchases/projects data. Use the historical emergency purchase list provided by the organization to the successful vendor to complete the following:
    - *Compare current inventory to historical emergency purchases/projects data, does inventory mirror historical purchases, and develop a variance list.*
- Deliverable 5—Compare OEM critical and recommended spares to current inventory and historical purchases. Use the OEM critical and recommended spares list provided by the organization to the successful vendor to complete the following:
    - *Work with the OEM to compare current inventory and historical purchases to critical and recommended spares, does inventory mirror historical purchases, and develop a variance list.*
- Deliverable 6—Validate on-hand inventory. This deliverable requires the vendor to be on-site. They must visually evaluate the existing inventory for spoilage, identify items fit for service, conduct belt survey, conduct gearbox survey, conduct bearing survey, conduct a pneumatic survey, identify obsolete parts, does inventory mirror historical purchases, and develop a variance list.
- Deliverable 7—Storeroom physical set-up. The vendor will assist in rearranging the storeroom in accordance with the organization's standards. Together you will develop storeroom design, physically move inventory, and assist in CMMS update.

Prepare a monthly progress report for review by the Project Team which will document all achievements in fulfilling the deliverables and flag any issues that require attention or reassessment. The report will include all output measures needed to quantify project deliverables.

Storeroom Organization for Productivity

Like any other business with productive operations, the storeroom must be organized in every sense. This means sizing the storage area appropriately for the level of business it will handle and equipping it to do so. The storage area must allow quick access and be free of clutter and debris. To minimize searching time, all locations need to be labeled. Handling equipment should be kept out of the main flow. The parts storage area needs to be secure in order to discourage theft. Encourage receipts/issues for inventory accuracy, and implement procedures to maintain this integrity around the clock.

An important aspect of storeroom operations is its correct and qualified staffing. To ensure productivity and personnel efficiency, the storeroom should be divided into work elements. However, limiting the number of work elements in a job increases productivity and reduces the time lost shifting between elements.

The realities of storeroom operation mandate that the staff are cross-trained since there will be a necessity to accomplish any function. Cross-training the storeroom staff will allow staff to move between different jobs on various days.

The Importance of Inventory Accuracy and Assurance

Inventory accuracy is a critical factor in the success of the storeroom. This accuracy is defined by having the right parts, in the right physical quantity, in a reliable state, in the right location, and correctly identified in the computerized maintenance management system (CMMS). There is some leeway with respect to small variances between the system and actual counts, such as with nuts and bolts. If the parts, location, or quantity are not correct when compared to the system, the location is in error and recorded as such when tabulating the inventory accuracy.

There are several reasons that inventory accuracy is important. Any inaccuracy can have consequences. If the actual inventory is below the system count, there is a risk of a stock-out situation causing parts to be unavailable and creating additional downtime. If the actual inventory is inaccurate, production operators and reliability technicians lose confidence in the inventory and CMMS. This may result in an additional loss of confidence in other aspects of these systems. Poor inventory accuracy leads to *gold rooms* (stock held outside the system in cabinets or other locations) when technicians lose confidence in the storeroom. These spare parts that are not tracked in the system create other issues that contribute to unreliability by not having the right parts at the right time.

Inventory assurance is equally as important. If your inventory is not usable, then it is not accurate. I have experienced several ways organizations handle inventory assurance and each has it pluses and minuses. Parts have expiration dates. A lot of people do not know or understand this. Expiration dates are a factor in ensuring that when you issue parts, you pull from the oldest and you stock the replacement parts in the back. This rotation of inventory will help prevent spoilage. Most organizations do not know that they should be maintaining the spare parts in their storeroom the same way they do the parts in production. The only way to curb infant mortality is to properly maintain your spares. Drives spoil because capacitors dry out, and seals fail in gearboxes, cylinders, and motors. Cylinders need to be stored vertically, not horizontality. All of these items need to be rotated, and if stored without lubricants will need proactive measures to prevent oxidation and corrosion. Turning shafts prevents false brinelling caused by vibration and exercises internal components. Many folks think there are a specific number of turns, or parts thereof, that are required. I practice rotating them enough to make one revolution on all internal components. This ensures that they are all exercised. The bottom line is that your proactive reliability strategy needs to include the equipment in your storeroom.

Inventory accuracy is not just a storeroom personnel concern. Only when all stakeholders in the storeroom inventory see this as a critical part of reliable operations will you achieve a reliable plant. Plant management, reliability, and operations personnel must understand the importance of this inventory since they all depend on its accuracy. To achieve a high level of inventory accuracy you must match parts receipts to repair and purchase orders and account for returned parts (previously issued) when they are returned to the storeroom. Capturing all information, serial numbers, or lots, upon receipt of each part will assist in its proper use and replacement. Inventory accuracy requires you to correctly account for your parts and their locations as they are put away, as they align with the work order, as they are issued and during cycle counts.

Why We Cycle Count

There are many variations to cycle counts that have been around for a long time. Parts can be counted in a variety of criteria, parts movement, value, or criticality. It can be done part to part at a different frequency.

There is more to cycle counting than how you handle out-of-tolerance counts. You need to have procedures. Counting is usually handled by a designated person whose main responsibility is to count. This may vary depending on storeroom volume. The counter receives a parts list, location, and criteria to carry out the count. Another important part of the count is to conduct it in real time, so the part's movement throughout the day will not impact its accuracy.

Once complete, the count will be entered into the system, without knowledge of the system quantity. If a mismatch exists within acceptable limits, the system is automatically updated. If the mismatch is not acceptable, the counter must do a recount. If after the recount there is still a mismatch the information is sent to an inventory control person to resolve. Once the inventory control person verifies the mismatch, an adjustment is made to the quantity of the inventory.

Proper cycle counting provides a reminder to the company that all employees are accountable for inventory accuracy. All inaccuracies will be investigated.

How to Inventory Correctly

You can increase storeroom productivity by properly putting away parts. Inventory slotting is the process of locating parts based on the usage, number in inventory, size and weight, and physical characteristics. The fast-moving parts need to be in the front of the room, allowing for quick access while the slow movers are near the back. If a gearbox is critical to the operation but is only used once a year, it should be in the rear of the storeroom. The opposite is true for a breather that is frequently used during a proactive reliability event.

When inventorying, it is important to determine the number and size of bins based on quantity and physical size of the parts to be stored. If a supply of parts requires multiple bins to hold their numbers, there will be a place to put them away. If they only require one bin, the location will depend on movement.

There are several ways to store and issue parts, and any of these may be employed depending on the characteristics and volume of the parts. In a storeroom with narrow aisles and dense storage, man-up vehicles might be the best option. Other parts may be stored best in a vertical lift module, which employs a carousel. An advantage of this method is that it directs parts quickly to the operator, reducing travel time. Less mechanized options are static shelves, small bins, and flow lanes.

Whichever storage method is used, the parts history must be analyzed to identify its movement. To justify the capital investment, deciding which method to use needs to be based on labor savings and productivity improvements. No matter which method is used to store and retrieve parts, each part must have a unique location and identifier in the CMMS. With a unique identifier, the CMMS will help streamline parts handling.

Bar Code Scanning

Most industries are using automatic identification methods to increase storeroom productivity, error elimination, and inventory accuracy. Using an automatic identification method is a cost-effective way to put parts away, cycle count, or pick parts.

In today's storeroom management, the need to write down part numbers and hunt for parts is dwindling. All information is sent via electronic means, eliminating errors caused by writing down long sequences of numbers and letters. Human beings make a mistake every 300 entries, while automated systems make a mistake every 50,000 entries. Using *radio frequency* (RF) to communicate the information via a voice-directed headset or a handheld scanner is today's answer. With the use of *personal digital assistants* (PDA), scanners and organizers have been merged. The advent of these technologies has assisted reliability technicians in getting information in real time and completing work orders in the field.

Proactive Maintenance ("PM") Kit Building

To increase the technician's productivity, the storeroom can provide tools, supplies, and parts for each preventive task. This is accomplished by the storeroom providing better assistance by building kits in advance of schedule.

The function of the storeroom is to provide the right materials to the reliability team to effect repairs. To level load the storeroom workload and provide better assistance to the reliability team, the storeroom can build reliability kits. These kits can be assembled in advance of the reliability tasks. Kitting can be accomplished by using a cart with multiple bins. The parts can be picked and placed in the bins. These bins are extensions of the storeroom so that the CMMS can track them and their location. Once picking is complete, the cart can be moved to a holding area for issue to the technicians.

Once the PM work order is scheduled, the technician will go to the kit holding area and retrieve the parts needed for his work. This provides a parts-retrieval process that is fast and helps the technician to remain productive. If a parts emergency arises and the part in the kit is the only one available, the CMMS can locate the part, remove it from the kit and issue it to the technician.

Parts Receipts and Parts Issues

Every part must be accompanied by a work order or issue record so it can be tracked by the CMMS. Parts receipts need to be tracked via the CMMS using a purchase work order, technician identification, and account code. Failing to follow established procedures will waste efforts and decrease efficiency, accuracy, and productivity. It is important that the reliability group embrace the CMMS and see it as a tool of success.

There are a lot of storerooms where the discipline of following the process fails due to the pressure of daily operations, especially during times of emergency. Additionally, a culture of "take what you need" may exist and be tough to overcome. With a lot of hard work, education, and accountability, these issues can be overcome. Remember, change takes time and changing a culture is not a one-time program but a continuous improvement process.

MEASURES AND KEY PERFORMANCE INDICATORS (KPI)

As managers, we can only expect what we inspect. To this end, you need a means to measure success. There are several indicators:

- Percentage of stock outs (number of stock outs/ total parts issued).
- Percentage of inactive inventory (parts inactive in the past year/total parts).
- Inventory accuracy (cycle count adjustment/ total cycle counts).
- Plant replacement ratio (parts inventory value/ plant replacement value).
- Growth rate in number of parts and numbers of suppliers.
- Parts to labor ratio (parts inventory value/ reliability labor cost).

Recording your indicators and measures over a period of time will result in the organization's benchmark. Using Edward Deming's wheel (*Plan, Do, Check, Adjust*), the process will continuously update the standards, resulting in productivity and costs savings. Only by tracking performance can you know if improvement efforts have achieved the results you expected and desired.

This gives you several areas to concentrate on to improve effectiveness and production in your storeroom. These enhancements are: organize the staff for efficiency, build reliability kits in advance to increase wrench time, adopt the idea that the storeroom is a reliability partner, be compulsive about inventory accuracy, be sure to record all spares moves, receipts, and issues, properly slot parts, use auto identification to streamline data entry, perform routine cycle counting and create, track and use KPIs.

Since all companies are unique, their storerooms will be, too. Be wary of consultants selling models, methods, or quick fixes. Each storeroom needs its own solution. There is no substitute for hard work and experience in accomplishing these efforts. When the storeroom follows a process, it is easy to see the increase in productivity in both the storeroom and the organization.

TECHNICIANS SKILLS

Improper installation is also driven by the skill sets of the technicians working within the organization. The overwhelmingly poor decision by most organizations to shortcut the hard work needed to produce skilled tradespeople is the largest reason the reliability teams lack the skills to properly install components. In conducting training for my organizations or my clients, I have experienced this first hand. When discussing the simple topic of fasteners, I like to ask the attendees how often they torque the fasteners when they install equipment. Without fail, the response is *rarely, if ever*. Now you are searching for your answer, and I bet most of you respond in kind. However, that answer is incorrect; the correct answer is *every time*. When you tighten a fastener, you are torquing it. Most people only consider it so if they apply a wrench designed to register the torque value. The truth is that both situations apply torque, but only one knows the amount. Precision reliability requires skilled tradespeople, and the only way to address improper installation caused by the lack of skilled tradespeople is to do the work and develop them. What I mean by develop them is to install a four or five-year apprenticeship, in collaboration with on-the-job development and mentoring by a real journeyman. An example of Reliability Department Career Progession is in Figure 4.1.

Operator Skills

The less intuitive reason for improper installation is the lack of a proper development program for the organization's operators. Here is yet another reason for unreliability. The reason I include operator development in the category of improper installation is because machinery that requires an operator is dependent on the operator. In order to run the operator-dependent equipment, there must be a *human machine interface* (HMI). The human is as integral to the operation as the equipment is. Improper development of these operators translates into the poor performance of the equipment. In a sense, the wrong or inadequately developed operator garners the same results as putting in the wrong part or installing it incorrectly. The mistake I have repeatedly seen across all industry is that most of this development is conducted on the job by another operator. Organizations that entrust this development to junior operators or a variety of different operators does so at its own peril. Repeatedly, I have seen operators who have months or even weeks of experience training the new operators. This is not development because neither operator is qualified to develop the other.

Another aspect of improper installation is the organization's desire to run the equipment beyond/outside its design. Most organizations do this because of a lack of understanding about the effects on reliability of operating

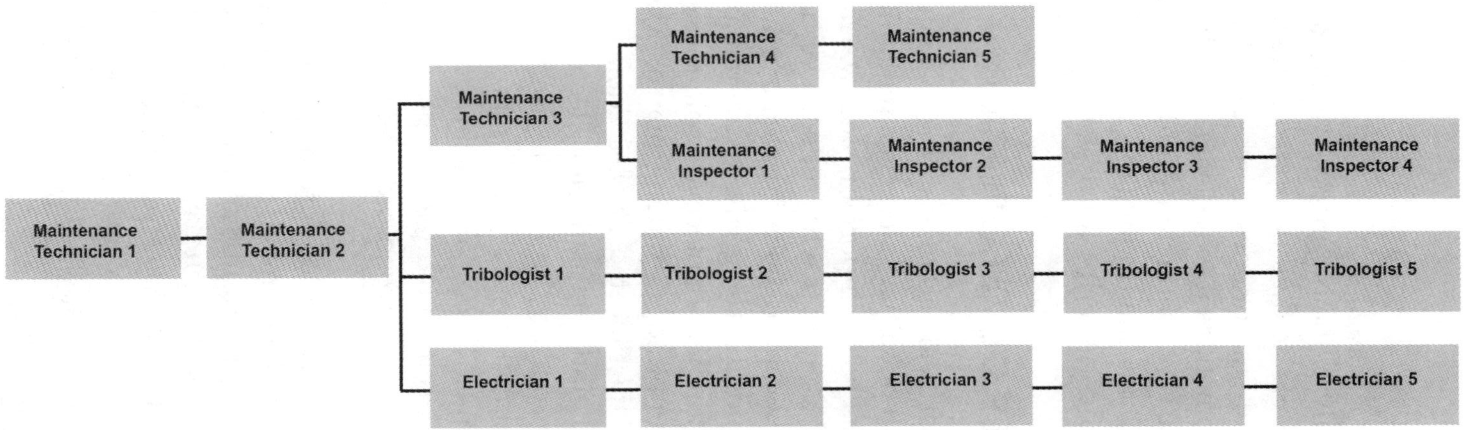

Figure 4.1 Reliability Department Career Progression.

equipment incorrectly. If the organization valued reliability, they would not operate their equipment outside its intended design. There are volumes of research that prove that operating equipment as it was designed to be operated will result in increased productivity and reduced repair costs. From an operator standpoint, the wrong operator doing the wrong thing is a recipe for disaster.

SKILLS DEVELOPMENT

Process training is a critical part of ensuring that your plant will operate to its full potential. However, reliability skills development is equally as important, and warrants the same consideration. No matter what industry or field you are in, everyone needs an organization that operates both successfully and smoothly. The goal is to operate problem free and to put development in place to avoid these issues. Figure 4.2 is an example of a training matrix. There are interactive software files located at *www.deathofreliability.com*. There you will find templates for electricians, mechanics, reliability inspectors, and tribologists. Because of aging equipment, human error, and different ways of life, problems are inevitable. To prepare for this environment, you need to arm yourself with a team of problem solvers and people who can look beyond today and act on what they see.

Training offers a crucial opportunity to grow the knowledge base of all team members, but many employers find the development opportunities expensive. Thus, they forego the development part and waste their money on training only. In reality, employers forego the development not because of expense, but because they lack the skilled professionals within their organizations to develop their

	Electrician Level 1	Qualified	Weight
	Must have a thorough understanding of basic electrical theory		
☐	Ohm's law	FALSE	5
☐	Parallel and Series Circuits	FALSE	5
☐	Must be able to perform equipment tests using appropriate instruments and methods		
	AC and DC voltage	FALSE	5
☐	Resistance	FALSE	5
☐	mA signal for 4-20mA and other control circuits	FALSE	5
☐	Must know basic hard wiring techniques	FALSE	5
	Must know motor control theory		
☐	AC	FALSE	6
☐	DC	FALSE	6
☐	Must be able to run conduit and wire	FALSE	3
☐	Must have ability to perform basic electrical Preventive Maintenance activities	FALSE	5
☐	Must have a high level of troubleshooting method	FALSE	10
☐	Must be able to diagnose electrical issues within limited time frame	FALSE	10
☐	Can design a plan of action to solve problem quickly	FALSE	10
☐	Is able to perform temporary fix, in order to get the machine back in production as quickly as possible and follow it up with a work order to ensure proper repairs are completed in a timely manner	FALSE	5
☐	Must have a high level of urgency when dealing with downtime issues	FALSE	5
☑	Must be able to read and understand basic electrical blueprints	TRUE	10
Percentage	10.00%	10	100

Figure 4.2 Training Matrix Example.

team. Notwithstanding the potential drawbacks, training and development offer both the organization and the team members advantages that make the cost and time a valuable investment.

All team members have weaknesses in their job skills. A development program allows employees to strengthen those skills that need improvement. Targeting individual needs instead of general training for the entire crew allows for better management of limited training dollars. A development program carries all team members to a higher level, so they all have comparable skills and knowledge. This helps reduce any weak links within the organization who rely heavily on others to complete basic work tasks. Providing the necessary training and development creates an overall knowledgeable staff with team members who can cover for each other as needed, work as a team member, or work independently without assistance and supervision from others.

Properly developed team members perform their jobs better than those who are not. They become more aware of safety practices and proper procedures for basic tasks. The development will also build employees' confidence because they have a stronger understanding of the industry and the responsibilities of their jobs. This confidence may push them to perform even better and think of new ideas that help them excel. Continuous development also keeps your team in the forefront of industry developments. Team members who are competent and on top of changing industry standards help your organization maintain a competitive advantage and remain a leader in its industry.

A structured training and development program ensures that your team has consistent experience and background knowledge. The consistency is particularly relevant for the organization's basic policies and procedures. All team members need to be aware of the expectations and procedures within the organization. This includes safety, discrimination, and administrative tasks. Putting all team members through regular development in these areas ensures that all staff members at least have exposure to the information. Team members with access to training and development programs have the advantage over those who do not have access.

Development programs create a supportive workplace. Team members may gain access to the development they wouldn't have otherwise known about or sought out themselves. Team members who feel appreciated and challenged through development opportunities may feel more satisfaction in their jobs, and this will increase retention.

RELIABILITY DEVELOPMENT

Reliability development offers the best way to build an organization of problem solvers. Your development program needs to focus on building a team that can maintain the equipment but also knows how to detect and identify issues. Any issue on the plant floor is an opportunity to improve performance and efficiency. Having a staff that communicates issues and solutions instead of hiding them proves that your organization's culture is promoting continuous improvement. Having this culture will be beneficial to the bottom line in both the short and long term.

Making the investment in skills development will enhance both organizational output and employee morale. Having a staff that is trained and can address any situation that arises will let you know that your team is well-rounded and is an improved operation.

Classroom training followed with practical experience could put an end to money being wasted on reliability courses that team members quickly forget. Organizations can spend tens of thousands of dollars on reliability training, but often, most of the knowledge gained in these training courses is never used in the plant. In fact, it is common for trainees to forget most of the lessons learned just weeks after the event. One way to help put an end to this phenomenon is to give classroom training in conjunction with practical experience.

It is a good idea to do your due diligence on all development provided in your plant. There is no justification for providing development that does not deliver on better reliability in the plant. There needs to be a plan with the end result in mind before any development takes place. Training for the sake of training is not a good business decision. If the plan does not include development once the training is complete, then save your money. There should be audits of the development plan to ensure the plan is being executed.

Management should also communicate to trainees that:

- results are expected from the training.
- support for the implementation of the course will be provided.
- there will be audits on the results.

The focus of a reliability department's development usually concentrates on lowering reliability costs and equipment reliability. A good example of this is alignment training. Mechanics usually perform an alignment after installing a new motor, gearbox, or pump, for example. The regular form of this training focuses on how to use the alignment tool. This approach achieves teaching the team how to use the tool, which is a necessary part of the tasks. A better approach would include a practical application of how to align these types of equipment. A comprehensive development plan has both a classroom and a practical aspect. A plant standard for alignment, along with an actual alignment specification for each piece of aligned equipment would ensure value in the development and enhancement to the plant's bottom line.

Most training classes teach theory about what to do and maybe give some hands-on classroom application. Some plants do a good job of training a few technicians in a classroom followed by practical application in the plant. However, most train several people, but never reinforce the training with practical application in the facility. The best approach is to include implementation of the theory taught in the class. The practical application should include safety, getting the right tools, scheduling the time on the equipment with operations, and what to do if the alignment standard cannot be achieved.

Great companies have structured training classes based on the above principles. Training courses in the essential care of equipment used to last for two or three days. In the new training approach, the courses last for three weeks. Despite taking much longer, the results showed that there are countless benefits to practicing standards that have just been taught. Including a practical aspect to training will help the team members understand the need for the standards and training and help them remember what was taught. Also, the participants get more motivated and participate in the training, because they can see how it applies directly to their daily work.

Making this training approach work will require a lot of planning. Scheduling the equipment, personnel, and other resources in a good plan will enhance the results of the training. Reliability leaders must put together a plan that includes: the working hours required, the development time needed, the development of condition-based routes, the time needed from the supervisors, planners, and personnel required for data entering into your computerized maintenance system, etc.

It is important that a baseline is established during the preparation period. Usually, it is easiest to take a few pictures in the plant. Issues such as lubrication, alignment, corrosion, balancing, and filtration can then be brought up during the course and visual examples displayed.

Spending your money on developing your reliability team's skills will not only increase plant output, it will also improve employee morale. When you have a properly trained staff armed with the skills necessary to perform their jobs, you will see a step change in your plant's reliability.

HIRE WELL

The best place to start in developing your team is to hire well. Hiring needs to be determined in reliability by qualified reliability leaders. At no time should you allow *Human Resources* (HR) to influence hiring of personnel that are not qualified. Hiring warm bodies is not the goal and is not considered hiring well. Filling positions for the sake of filling them does not strengthen your team or improve your proactive strategy. The hiring of unskilled handymen into journey-level positions is diminishing the pool of qualified craftspeople. Your search needs to focus on people who are eager to learn and to continuously improve. To improve your team's performance, you need team members who are engaged. If your team members are striving to learn more about their industry and organization, they will perform better in their jobs and see them as their careers. Development is vital to any employer's ability to retain its team members, build loyalty, and harness the full value of its team.

It is imperative to help your team members understand how their contributions to the process help the overall function. Not everyone can see how they fit into the big picture. When an employee has a job with a limited scope, they can feel like their efforts are not important. Through development, the employer can help the team members understand how they fit into the organization's goals, mission, and structure. This will lead to more highly motivated team members who understand how they contribute to the overall success of the organization.

ENGAGING THE TEAM

Engaging the team in the improvement process will capitalize on one of the eight wastes identified in lean manufacturing, intellect. The employees know more about how to improve the process than anyone else. Companies hire outside consultants to point out opportunities, but none of these consultants has the insight of the person doing the job. The goal is to develop your employees to see the opportunities and give them a process to bring these ideas to the proper people. Most of the time, all the employees' need to reach their full potential are tools, education, or organization. Enabling your teams to help get things right will improve output and quality. If people are happier in their work, the chance that they will have success increases exponentially.

TRAINING

Training for the sake of training, or to "tick a box," serves no real purpose. Evaluating the skills of each member of your team is necessary in order to build a development plan for the individual. People do not join companies for a job, but for a career. To increase performance and retention, you must provide them with opportunity for advancement. These opportunities must be real. Prospects that are a theory will lead to disenfranchisement. Helping team members realize their goals will motivate them to perform at a higher level. Education will open their eyes to the business and can result in leadership skills for the future.

It is important to send the right message to your team. The investment an employer makes in their team members lets them know they are valued. The investment tells them they are a long-term solution to the company's success and that the organization values continuous progress. This investment creates loyalty and enthusiasm in the team and strengthens retention. The reason most training fails to achieve the desired results is that training alone cannot achieve the results. Training is the reason development fails. Before sending your team off to training, you need a development plan designed for everyone that is focused on the objectives.

TRANSFER OF SKILLS

To transfer the skills desired and addressed by the development sessions, it is imperative that you do the following things before starting the development. Number one is to make sure you need the development. This requires a skills assessment to determine each employee's real need. Make sure that the opportunity you are after or the problem you are trying to solve can be answered by development. If the employee is not performing at the expected level, determine whether or not you have provided the necessary development to do the job. Consider if the job is a good fit—does the employee have the aptitude for the position? If the answer is no, then development is not the answer. People are either willing but not able, not willing but able, or willing and able. Until you understand where the employee sits, sending them to training should wait. Provide context to the team members. Make sure they know why they need a new skill, enhancement of their current skills, or other reasons the development is being offered. Employees need to see the connection between their jobs and the development. When they understand this connection, they will be motivated to find the relevant information in development. Focus your training and development on the relevant skills you need the team to have. One-size-fits-all training programs designed to provide generic information are not what skilled professionals need. Customized approaches designed to meet specific objectives with follow-up development will offer the greatest return on your investment. Design your approach so it is measurable and focused on specific outcomes. Tying the development objectives directly to the employee's job and skills objectives will motivate the employee to attain the promised objectives. Keep the message clear and show employees how the development will enrich their jobs.

Improper installation can be compared to a milking stool—it has three legs that keep it from falling. The three legs are: design, stores, and training, and if any one of these is not present the person sitting on the stool must try to balance while also trying to produce. If you have three sturdy legs, the person/organization has the foundation to perform at its peak in this area. Which organization do you want to work for and what type of leader do you want to be? ∎

CHAPTER 5

What is a Reliability Professional?

A RELIABILITY PROFESSIONAL

As I mentioned in the Introduction, an ideal reliability professional has a variety of attributes: a lifetime of real-world experience, a master or journeyman's license, trade certifications, professional licenses, and a professional reputation of success. The work experiences that enable them to develop and lead a reliability team include stints as a combination of the following: maintenance person, planner, supervisor, reliability engineer, and maintenance or reliability manager, depending on the level of their position. In addition to this level of practical real-world experience, they possess an advanced education with finance, development, change, and leadership concentrations. A Navy SEAL friend of mine use to say that "they don't bring SEALS in to negotiate, they don't bring them in to be fair, they don't bring them in for any other reason but to win. It's not a self-help group—they are there to take over and tilt the odds back in favor of America." Now I am not equating a reliability professional with a Navy SEAL, but there are similarities in the philosophy. A reliability professional does not negotiate; they are not fair, they take over and win. In the case of reliability, this unfolds as a dogged effort to transform an organization from production-centric to productivity-centric. Production cannot and will not happen without reliability. It takes a hard-charging reliability professional to move this needle. These professionals operate and dwell in the space between reality and fantasy. The reality is that most organizations do not value maintenance but want reliability. When a true reliability professional leads an organization down the path to reliability, some might think it is fantasy, but it is only a fantasy if the reliability leaders are engineers or consultants who are not qualified by experience to lead the efforts.

Job Description

The first step in finding a real-world true reliability leader is to write a good job description. Specify the qualifications the organization is looking for. I have reviewed thousands of job offerings from all industries searching for a Maintenance Manager. These offerings list having a mechanical or electrical engineering degree as their first qualification. This tells me that the organization does not understand or value reliability and does not have a robust reliability approach. They feel that a four-year degree qualifies an individual for the position when the reality is that it qualifies them for nothing. The first and most important qualification for any reliability leader to have, especially the Reliability Manager, is a journeyman-level skilled trade. I have found that an electrical or mechanical emphasis is relevant here. It is absolutely a requirement for a reliability leader to have done maintenance. Without this key experience, they will never truly understand how to lead the team.

IDENTIFYING A RELIABILITY PROFESSIONAL

To identify a reliability professional all you must do is ask them "how exactly do you go from run-to-failure to proactive reliability?" It is important that the senior leader conducting the interview be a qualified reliability professional who will know if the answers are right. Prospective reliability professionals should be able to explain the detailed steps necessary to achieve the results. They can give examples of hands-on, real-world experience doing these efforts, not regurgitate something they read or a course they took. They can assess an organization's level of reliability leadership by assessing if they have a mature maintenance and reliability program. Does their CMMS work? Is their *maintenance, repairs* and *operations* (MRO) functional? Do they have a work management process in place and do they have a reliability-focused proactive approach (programs addressing lubrication, contamination, and improper installation)? Companies who entrust their new efforts to the same individuals, without the qualifications, repeat the same failed approaches over and over. The fact is, despite a lot of excuses, if the tenured staff had the necessary reliability skills, the organization would have reliable operations. When I hear company leaders say that "it is not necessary to have done maintenance to lead it," I know they do not understand or value maintenance and reliability. No one can lead maintenance and reliability unless they have done it.

Create

You do not gain a competitive advantage by copying others. True leaders create and do not need to copy. Each organization is its own entity, and you cannot copy your way to success. Reliability leadership needs people who can create on their feet using years of experience and education. The respect of the reliability personnel is necessary, and that comes from the shared experience of doing the jobs you are asking them to do. Reliability is not a theory, but the result of having held each maintenance and reliability position and the ability to train and develop the personnel needed to drive success.

The organization of the reliability department is critical to its success. The most important aspect of this organization is that the Maintenance or Reliability Manager must report directly to the General/Plant Manager. Any other iteration does not provide a strong enough structure for an organization to truly realize the competitive advantage that a robust reliability department can provide.

The structure of the department is important to a strong reliability organization. The minimum structure needs to include three branches: planning, execution, and reliability. There is more to this structure than just having it. It is imperative that the branches focus on their responsibility. If the department cannot focus the branches to concentrate on their duties, the results will be mediocre. The current trend in business is for everyone to do everything and it is this failed approach that leads to nothing being accomplished. If everyone is responsible for the results, then no one is.

RELIABILITY LEADERS

Are leaders in reliability a good idea? This is like Mahatma Gandhi's answer when asked what he thought of Western civilization. He replied that he felt it was a good idea. This is the same thought I have when considering management in reliability.

In my experience, reliability leaders are regularly promoted because they are good technicians. I often refer to this as the wave of a "magic wand." The ability to turn a technician into a leader in this manner is beyond me. How can you transform a person who has never had a leadership job into a leader? The senior managers who make these decisions believe that it can be done without a job description, coaching, or development. This transformation takes place at quitting time on Friday and the person returns on Monday a leader. This "magic wand" must be the greatest tool available on the market.

Other examples of these magically born leaders are young engineers being placed in supervisory positions or senior engineers being selected as reliability managers. I would like to share my thoughts on leadership in reliability, especially for those who have encountered the "wand."

WHAT MAKES YOU A LEADER?

Leaders do not have subordinates, at least not when they are leading. Many leaders do have subordinates, but only because they are also managers. But when they want to lead, they must give up formal control, because to lead is to have followers, and following is always a voluntary activity. There is a significant difference between people doing what they are told, and following. If you do not get this, then you are not a leader.

A Leader Must Know What to Lead Towards

There are a lot of theories when it comes to leadership. The whole idea of leadership falls apart if the leader does not know what beliefs to lead towards. In *The 7 Habits of Highly Effective People*, Stephen Covey said to "begin with the end in mind." The leader must know the long-term purpose of reliability in their organization. The organization must document and communicate its beliefs for the management of reliability. If they do not, they have set their leaders up to fail. Former New York City Mayor Rudolph Giuliani was asked what he thought was the most important thing to know as a leader. His answer was to know "what you believe in before you start leading others." As the reliability manager, you must set up a clearly defined "place to go" for your reliability efforts. These efforts revolve around preventive maintenance, planning and scheduling, root cause problem elimination, spare parts management, design for reliability, skills improvement, and other tools. To lead in reliability, you must have real hands-on experience and the supporting education. It is nowhere near enough to have held a leadership position to truly lead reliability efforts. It is a specialty skill that there is no shortcut for. Only hard work and relevant experience will result in success. It is the difference between good and great.

Where to Start

When considering where to start you need to work to clarify the organization's efforts. You might state "you will store lubricants properly" or in spare parts management, "you will have an inventory assurance, not accuracy, of 95 percent."

Regardless of whether you are a corporate executive, plant or maintenance manager, planner, or a supervisor, ask yourself if you have a set of values to lead towards. If not, I suggest you make a list jointly with your organization before setting out on this journey.

The Role of a Reliability Leader

The next part of reliability leadership is recognizing the role of a leader. A lot of people think that the role of a leader is to tell others what to do while others believe it is to motivate and encourage people. I think this is misguided. The role of a reliability leader is to simply get others to follow them. As I said earlier in this book, a title does not make you a leader—having followers does. Following is a voluntary act so you must inspire people to want to follow. You cannot lead from an office; leading is accomplished by encouragement, motivation, group discussion, and much more with a goal of accomplishing the company's objectives. We will discuss what is required to develop these transformation abilities in people. To ensure the leader is aligned with the company objectives, these objectives must be well defined.

Followers

How do reliability leaders get the reliability team to follow them? This is an art that few master, but can be improved upon. In my experience, a journeyman-skilled reliability team will only follow a leader with journeyman-level skills. Again, some of you might be thinking that your team follows you and you do not have a skilled trades background. This just tells me you do not have a team comprised of journeymen.

To get people to follow you, you must continuously build trust and processes that enable them to perform at their best. People can never be more effective than the systems they work in. When asked "how much time does your reliability planner spend planning reliability work?" sixty percent of those surveyed stated that less than 30% of their time is spent on planning. Twenty-six percent said they spend less than 10% of their time planning. In a follow-up survey, the top three reasons given for these percentages were: emotional priorities, too many do-it-now jobs due to breakdowns, and operations not supporting the planning process.

Leaders must set up roles and responsibilities for planners and others involved in the planning and scheduling process to make sure the process is followed. These people need to work with operations, engineering, and stores to agree upon work priorities, schedule cutoff times, and identify critical equipment, spares, and much more. Your chance of getting people to follow you drastically increases if the enablers are instituted plant wide.

Setting the Example

A leader must realize that people do what they see being done, not what they are told to do. It is crucial to "walk the walk" by sticking to your plans and following up on what you say. To go back to what I said earlier about what constitutes a reliability professional, to "walk the walk" you must have done the work. The "magic wand" will never replace hands-on experience and the necessary skills to deliver results through real leadership.

If you are a leader that shows up to meetings late, you will find it difficult to instill good scheduling practices in your team. Leaders should not expect quality work if they are constantly asking for results in last-minute and unreasonable time frames. A leader should not expect quality results if their team members are not properly trained or lack the financial resources to effect quality repair, or they are not given enough time to complete the tasks, and no standards or expectations have been communicated.

TRUE LEADER

A true reliability leader makes sure that all improvements have substance to them by ensuring that there is a solid plan with costs and benefits considered before rolling it out to the organization. All too often plants sign up for a project-of-the-month, only to replace it with a new project every few months. Hiring a consultant to lead your reliability efforts is an example of a project-of-the-month. All consultants contort the assessments they do to meet their offering and needs. When you decide to use a consultant, it should be limited to a strategic, very focused task or tasks, and never a one-size-fits-all broad stroke "model."

While visiting a plant a few years ago, I mentioned that reliability improvements should go on forever. It is a continuous process. A tradesperson in the audience said, "In this plant, forever is six weeks, and the yield for reliability enhancements in six weeks will most likely be honestly small." I was not sure what he meant by his statement. He further explained, "every time we start improvement efforts that are announced to last forever, the average life of a new initiative here is about six weeks. Therefore, forever here means six weeks." Despite the fact that he was partly joking, he was exactly right when it came to the plant's history.

The Gap

I remember someone once said to me, "There is a small gap between inexperience and knowledge. There is an enormous canyon, however, between knowing and doing." This resonated with me because it is exactly the same difference between a consultant and a reliability professional. A consultant is all about knowing the theory and a reliability professional is all about knowing *how to do it.* It applies here because a reliability professional knows how to build the bridge over this canyon and get lasting results because they have done it.

As a reliability professional, you know what to do. You know why you should do it. And more importantly, you know how to do it. Most people don't change, either as individuals or in the way they run a reliability program. The problem is that most improvement efforts assume that all you need to do is explain to the team that things will be better and then tell them to do it. Does that work in real life? For example, people are told, "Smoking and excessive drinking are really bad for you." "Oh, well then, everyone will quit right away." If this were that easy, we would have no smokers, alcoholics, drug addicts, over-drafted credit cards or gamblers.

I have spent my life developing people in scheduling, planning, root cause analysis, proactive reliability, and storeroom management. A request for consulting and development has always been a pleasant part of my professional life, but it is also important for me to ask any potential client, "What is your development plan once my work is complete?" Sadly, the answer is normally nothing. Their expectation is that the development or consulting is the permanent fix to their issues. They believe that by training people to perform root cause analysis, they will solve all the plant's woes. By consulting on reliability, their team will suddenly be able to lead the implementation of a reliability strategy. Training is the number one reason development fails. If you plan to train your way to reliability, you should save your money. All you will be doing is paying for the next flavor of the month.

Awareness training is one of the first steps, but you obviously need to remember what the desired end results are. Execution is the key, not knowledge by itself. True reliability professionals know how to execute where others, sadly the growing majority, do not have a clue. Every day organizations are setting up engineers and production employees for failure by placing them in leadership positions in reliability. Leadership in these areas is not something you can learn in a book, classroom, or seminar. It requires years of hands-on experience and development with "boots on the ground." Most organizations are looking for a quick fix and if that is the objective you are better off saving your money. You cannot copy or cheat your way to reliability; you must do the work.

How to Motivate Reliability People

Do you believe there should be a different strategy to motivate your reliability team than the one you use to motivate other plant personnel? My answer is no. A good thing to remember is that people are people, whether they have funny nicknames taped to their hats, are salaried, or union members. The main question at a lot of plants is how to motivate people.

A cheap hat or pizza lunch has never, by itself, inspired anyone to do a better job. The top prize in un-motivating motivation goes to an organization a friend of mine worked for. During Christmas time, all the employees were given Lifesavers candy with a note saying "you are a lifesaver." How is that for motivation? The organization was a hospital, but still, it did not achieve the results it intended.

How do leaders motivate their people? The first thing you must realize is that there are few people who can inspire people en masse through a speech. With the exceptions of Ronald Reagan, Winston Churchill, Mahatma Gandhi, and Nelson Mandela, leaders need to rely on motivating their employees one-on-one. True leaders need to take the time to listen, challenge approaches, and discuss ideas with each direct report. For most people, pride in their job is an incentive. Leaders should provide honest feedback to their employees using skillful diplomacy. For the recognition to have merit, the leader must have the employee's respect. To accomplish this the leader needs to have earned it.

WORKMANSHIP

Pride in and recognition of their workmanship is a better incentive than gifts in motivating people, even though money does have its place in any incentive program. There are many people dedicated to dangerous and relatively low-paying jobs, like soldiers, which makes it hard to believe that money is the number one factor in getting people motivated. With that in mind, set up the right work processes and closely follow up on the progress to achieve change.

As I have explained previously, to effectively get people to do what you need them to do, you need systems and procedures to support them. People are only as effective as these systems and you need to start by building these systems and procedures. However, as a reliability leader, if you are going to get your team to do what you need, you will need to use your experience and earned respect, not psychology and diplomacy.

Consider the following situation: you tell your significant other that their face is so ugly that it could stop a clock. This would lead to a fight and possibly other repercussions. Now let's take another approach. Tell your significant other that when you look at them all time stands still. This would have the opposite result. If you think about what you said, it was the same thing. So, spending the time getting to know and understand your team members and what motivates each one individually is a requirement of leadership.

When charged with leading people, how you talk to them is important. This same principle applies in a plant, and if you are not doing this already, you will see improvement in performance if you are mindful of how you speak to people.

PEOPLE ARE NOT YOUR MOST VALUABLE RESOURCE

You are always told that people are your most valuable resource. This can't be true. When I visit plants, I hear the same story. "The supervisors tell me that they could send some of their reliability team home and it would not change anything." This is because this portion of people do not contribute anything useful. The truth is that people are not the most valuable resource, the *right* people are.

It is, therefore, as Jim Collins mentions in his book *Good to Great*, utterly and completely vital for any plant to "get the right people on the bus." If you have a weak corporate reliability leadership group, reliability manager, supervisor, operations coordinator, planner—you name it—you cannot be successful in your efforts to improve reliability. Senior leaders and plant managers must do something about poor performance. If they don't, they will run the risk of losing their own jobs, because the same is true for a corporate vice president.

ORGANIZATIONAL STRUCTURE

Before we can get the right people on the bus, we must build the bus. The basic structure of a reliability department needs three functional groups at a minimum. The three groups are:

1. Planning
2. Execution
3. Reliability

This is illustrated in Figure 5.1. It is key to focus each functional group on their tasks. If you have everyone doing everything, you have no one doing anything. The bottom line is that you must have clear expectations for each team member and communication between the groups to be successful.

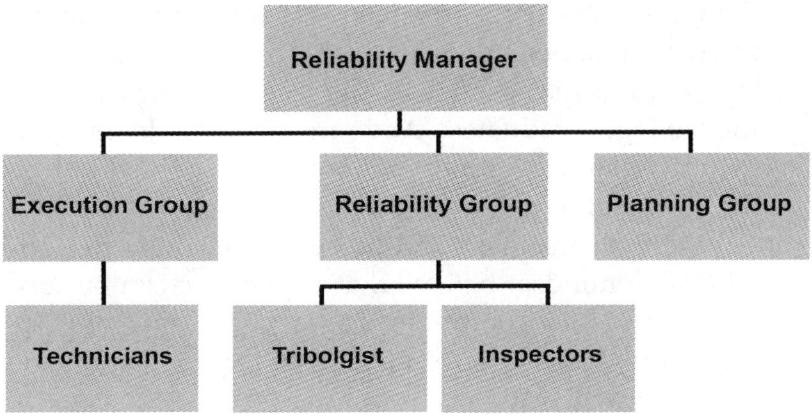

Figure 5.1 Sample Organization Chart.

Planning Group

The first group we will discuss is the planning group. We will direct our attention to the requirements for selecting a planner a little later in this chapter, but the focus here is planning your work. The planner needs to ensure that each job is set up for success. The job must contain safety, drawings, procedures, parts, etc. Planners cannot be assigned any other duties. They do not coordinate contractors, cover for supervisors, or perform any other duties not directly related to the plan generation. One planner effectively producing a plan can reduce the need for additional manpower when compared to organizations that do not have a planning group. This can help organizations that are short on manpower and cannot get it because of budget constraints or the available pool of skilled tradespeople in the area.

As we will discuss a little later, it is imperative that the planner be a skilled tradesperson. This is another leadership position that cannot be filled by inexperienced or young college graduates. Any organization looking to shortcut its planning group and the efforts carried out by it will do so at the risk of its own demise. Here is another place where a consultant and those management personnel with this mindset believe that experience can be substituted with fancy computer programs or pages of pretty procedures. Speaking from a lot of experience, this is wrong and it will lead only to failure.

Execution Group

The second group is the execution group. It is this team that executes "work," or the plan. Once the plan is finalized it is passed to the execution team leader/supervisor, who will assign personnel based on skills and available resources. The focus of this team is to complete the plan and if there are break-in requests, they adjust the plan and execute the break-in work. The ability to prioritize the plan and any break-in work is a must for this team leader. Later in this chapter we will list the requirements for this team leader, but knowledge of the plant is vital to understanding criticality and priority for plan execution and adjustments when necessary. I use the priority guidelines shown in Figure 5.2.

The difficulty here is changing your culture. Most organizations have a reactive reliability approach and they reward the team for its ability to react (firefighting). The execution team will ride in on its white horse and quickly patch the equipment and return it to production. There is no thought to a proper repair, or understanding the root

Reliability Priority Matrix						
Time to Unreliability	Priority					
Less than 2 days	7	3	2	1	1	1
2 to 7 days	7	4	3	2	1	1
7 to 14 days	7	5	4	3	2	1
14 to 30 days	7	6	5	4	3	2
More than 30 days	7	6	6	5	4	4
Health & Safety	No Effect	Concern Only	First Aid	Lost Time Injury	Multiple Injuries	Fatality
Environment	No Effect	Concern Only	Internal iusse	Loss of compliance - Department	Loss of compliance - Plant	Massive Impact
Costs	No Effect	<$5K	$5K - $50K	$50K - $1MM	$1MM - $5MM	<$5MM
Relationships	No Effect	Concern Only	Minor Impact	Loss of reputation - Local	Loss of reputation - National	Business Loss

Targets							
Priority	7	6	5	4	3	2	1
No Shutdown Required	Reject	More than 30 Days	14 to 30 Days	7 to 14 Days	2 to 7 Days	2 days	Immediate 24/7
Shutdown Required	Reject	When people are available	Next 2 to 3 shutdowns	Next scheduled shutdown	First opportunity	Shutdown within 48 hours	Immediate 24/7

Figure 5.2 Reliability Priority Matrix.

cause and eliminating it. In most plants, the execution team responds to radio calls from anyone who has a radio or happens to walk by them and ask for assistance. The education that is required to help the plant understand that this methodology hurts the plant and its productivity takes time and support by all levels. Most organizations are extremely shortsighted, despite boasting of the opposite. The goal of the execution group is to execute the plan, not react to emotional demands by everyone in the plant.

Reliability Group

The third and final group is the reliability group. It is this team's job to look to the future. They are the eyes and ears of the proactive reliability strategy. Normally this group consists of "tribologists" (lube techs) and reliability inspectors. Training, development, and certification of these team members is a requirement and without it you have nothing more than glorified execution group team members. There are too many organizations that see their tribologist as entry level personnel. In fact, if the organization employs this strategy, it is not properly structured or manned. Your tribologist is the phlebotomist of your equipment. As in the medical profession, these team members will determine the accuracy of your proactive efforts. Lubrication is the life-blood of the equipment, and sampling and testing it will drive your proactive efforts. Again, your reliability inspectors are the ears and eyes of the reliability team. They are conducting the physical exam of the equipment using a variety of diagnostic tools. Like the medical profession, these team members interpret physical signs and test results to direct proactive efforts. Your tribologists and inspectors are not an extension of the execution team. Their efforts must be the priority of your department and organization. Only when your organization embraces proactive reliability will it realize reliability.

These three teams need to focus their efforts in their assigned responsibility and the leaders in each area must protect their team's tasking. You should not allow your team's resources to be wasted by not accomplishing its planned work. It is not acceptable to assign these team members to any tasks other than their assigned proactive efforts. All too often, an organization will suspend proactive work in favor of reactive work. You will know you have a proactive organization when your entire organization prioritizes the reliability team above all else. Yes, above even production. Most organizations are so short sighted that they cannot make the leap from "production at all costs" thinking. The truth is, if you prioritize proactive efforts over even production, you will meet your production needs because you have a reliable operation. Reassigning your reliability team should always be the last resort. You should shut down a line before you fall back to your old ways and rob this resource. This is one of the toughest obstacles to overcome, because most leaders cannot make the leap from doing what they have always done. The fact is, if you make this leap, you will realize the best production your plant is capable of. Now that you have built the bus, you need to fill the seats. The next few paragraphs detail the requirements necessary to build a qualified leadership team for your reliability efforts and put the right people on the bus.

Reliability Manager

What do qualified reliability leaders look like? At a minimum, Reliability Managers have more than 20 years of experience in reliability. They possess a journeyman-level skill and have worked as a reliability planner and frontline supervisor. Depending on your company objectives and size of the reliability workforce, the Reliability Manager should have advanced education in managerial finance and leadership. To round out their qualifications and ability to successfully lead reliability efforts they should be a licensed Project Management Professional, Certified Maintenance and Reliability Professional, and certified in proactive reliability approaches (lubrication, contamination, and improper installation).

Reliability Engineer

At a minimum, Reliability Engineers have more than ten years of experience in reliability. They possess a journeyman-level skill and have worked as a reliability planner and frontline supervisor. Depending on your company objectives and size of the reliability workforce, the Reliability Engineer should have advanced certifications in vibration, ultrasonic, and thermography. To round out their qualifications and ability to successfully lead reliability efforts they should be a licensed Project Management Professional, Certified Maintenance and Reliability Professional, and certified in proactive reliability approaches (lubrication, contamination, and improper installation).

Reliability Planner

At a minimum, Reliability Planners have more than 10 years of experience in reliability. They possess a journeyman-level skill. To round out their qualifications and ability to successfully lead reliability efforts they should be a licensed Project Management Professional, Certified Maintenance and Reliability Professional, and certified in proactive reliability approaches (lubrication, contamination, and improper installation).

Reliability Supervisor

At a minimum, Reliability Supervisors have more than five years of experience in reliability. They possess a journeyman-level skill. To round out their qualifications and ability to successfully lead reliability efforts they should be a Certified Maintenance and Reliability Professional and certified in proactive reliability approaches (lubrication, contamination, and improper installation).

THIS IS IMPORTANT!

This is so important that it bears repeating: *The propensity to hire engineers to fill reliability leadership positions is one of the most significant reasons reliability is failing across the world.* Engineers, both young and old, are not qualified to lead reliability based solely on degree or title. This is also true of production personnel and consultants. Organizations are no longer willing to spend the time or money necessary to build a strong organization. They are looking for a quick fix and not a real solution. There is no shortcut or easy way to accomplish true reliability. It takes hard work and knowledge, which is not taught in school or offered by a consultant; it takes a lifetime of efforts and experiences.

Why to Choose Craftspeople

The heart and soul of any reliability team is the skilled craftspeople. Here is another place the quick-fix approach has failed the maintenance and reliability community. You hear over and over that the reason the skilled craftsman workforce is declining is due to the lack of interest by the next generation in undertaking the hard work associated with reliability. It is true that a small portion of the decline is being driven by today's youth not wanting to work as craftspeople, choosing instead to sit at a desk and manipulate a keyboard. The larger contributing root cause of the decline in skilled craftsmen is the latest propensity to hire engineers, young and old, as leaders. Engineers have not earned their way to a reliability leadership position and do not understand or value the skills of a journeyman. Since they followed the shortcut method to leadership, they look for the same quick fix when addressing their team's skills gap. The lack of reliability professionals, as described above, has caused the decline.

How does leadership and the quick-fix approach befalling industry today create the growing gap in skilled craftsman? The answer is easy. Engineers do not understand the importance of a craftsman apprenticeship. They feel that the education necessary to be a skilled craftsman can be "fast-tracked," and it is unnecessary to spend the time and effort necessary to invest in true skilled trades. One reason they believe this is because they never put in the work and effort that is required to complete a trade apprenticeship.

An apprenticeship program is a combination of course and hands-on practical education. This program is instructed by tradesmen who have spent their lives building on the skills necessary and contributing to the body of knowledge that has culminated in the program they must complete. It is impossible for a college or university to transfer this level of skill, largely because a college degree does not qualify anyone for anything. Despite this, many try and fail. Companies are desperate to shortcut the process and save money; they do not care if the skills are not transferred. This approach factually costs a lot more money than the savings they believe they obtained by taking the shortcut. Most companies espouse that their first value is a dedication to safety. However, these shortcuts place their employees in danger, because the necessary skills cannot be achieved through shortcuts or taught in a college. Once again, I will say that there is no shortcut or easy way out when it comes to reliability. The only way to get true results is for your organization to roll up its sleeves and get dirty. If you have sold or are attempting to sell yourselves on this approach, you should save your money. All you will accomplish is putting your employees in danger and your equipment will continue to degrade.

WHY THE PURSUIT OF QUICK FIXES?

Why do so many organizations look for quick fixes? It starts with the false belief that engineers or production supervisors can lead reliability. The belief is that you do not have to have done reliability to lead it. It is further compounded by the belief that to help your engineers or production supervisors become reliability leaders you should employ consultants. The greatest number of consultants are engineers or production personnel and they have even less experience than the company's own staff. They have developed or butchered someone else's work to sell upper management on their own worth. Another error most organizations make is when upper management hires outside consultants to "fix" their maintenance and reliability efforts. The issue here is that no one in upper management knows reliability. In my doctoral studies, I found that less than four percent of upper management has any exposure to reliability and these are the people charged with passing down the edicts that shortcut leadership, training, etc., things that place all the employees involved in harm's way. Even worse is that these upper managers do not realize what they don't know. They believe that any manager can lead these efforts. This cannot be further from the truth; leadership in reliability is a specific skill and it requires years of real- world, hands-on experience and education to be qualified to lead reliability. Until organizations understand this, they will be placing their employees in danger.

BRINGING IT ALL TOGETHER

To summarize, improving plant reliability to the point of perfection is a long journey with many milestones. But plant reliability and any other improvement efforts must start with the following actions: get the right bus (structure), get the right people in the right spots, and create a clear vision for the plant to strive for. Managers must be leaders—they cannot just manage—and leadership must create a work system that allows people to be effective. This same structure needs to be duplicated throughout the organization. It does not work without continuity at all levels. If your organization has a reliability structure at the corporate level, you must select a leader and structure that supports your plants. The only qualified leader for this organization is a qualified reliability professional. It is necessary that the reliability leaders, at every level, have hands-on experience doing reliability. Without it, your organization will fail.

There is no shortcut or quick fix for building the right structure and hiring the right people. If your organization attempts to shortcut the above structure and manpower, you will fail. The time you will waste, the profitability you will lose, and the increased costs to the organization will be more than you will pay for doing it the right way.

CHAPTER 6

Leadership Style

MY DOCTORAL RESEARCH

This book will touch on the lack of representation of true reliability management personnel on the *Top Management Team* (TMT). This chapter includes a lot of the research that I have relied on in writing this book. I think it is important for us to understand what is meant by "true leadership." I ask that you bear with me and listen to what it is that I am saying as it is related to leadership. Most of us do not get to spend 14 years of our lives researching leadership, so this summary will catch you up on that research in just a few minutes. It is impossible to lead your organization in its transformation with a transactional leadership style.

The fact that reliability can account for over 40% of an organization's budget and manpower is not weighed when selecting the TMT. Reliability is usually represented in the TMT by operational managers. The focus of operational managers is production. The mindset of a reliability manager is production capacity. In most organizations reliability is not considered a part of the organization's strategic planning or competitive strength equation. Most organizations' TMT spends less than three or four percent of its time considering reliability issues. When they do they concentrate on the cost and not the management of this key part of the company's overall success. When a company is looking for waste within its operation, the TMT does not realize that its creativity would be better spent concentrating on reliability problems and how it is operating its equipment. Organizational reports focus on profits while reliability is lost in the maze of other costs. Companies look to the obvious: airlines look to the number of seats sold and schedule compliance, hotels look to the number of rooms sold each night, schools look to enrollment, and manufacturing firms look to units produced. Buried in all of this is the cost to keep the planes flying and on schedule, the rooms suitable for occupancy, the environment suitable for the students to learn in, and the equipment necessary to produce the units defect free. The value-added potential of true reliability management can have a significant influence on the bottom line. This is where the lack of the TMT reliability background is detrimental to the company's overall effectiveness. There is more to reliability management than just costs, and that is where representation by an operations manager falls short. Despite their overwhelming qualifications, why do corporations fail to view reliability management as a resource for the Top Management Team?

Top Management Team

I would propose the addition of a reliability background to the functional track currently considered when selecting a corporation's TMT. I hope to remove the obstacles that currently exist which prevent reliability professionals from ascending to Top Management Team positions within corporations. I believe that adding this functional background to the composition of the TMT will strengthen the organization and result in greater profitability. Without it, the death of reliability is certain.

During my research into leadership style and its effects on company profitability for my dissertation, I found that the historical concepts of transactional and transformational leadership began with James McGregor Burns. He was the first to propose that followers will strive for goals that characterize their values and motivation. He defined transactional leadership as an exchange between followers and leaders of one thing for another. Transformational leadership focuses on leadership between followers and leaders to help raise each other to a higher level of morality and motivation. Bernard M. Bass believed that transformational leadership was more powerful than transactional leadership. There is a considerable body of literature that establishes a positive relationship between transformational leadership and company performance (profitability), follower outlooks (i.e., job satisfaction, organizational commitment), and performance (i.e., profitability) at the individual level.

There have been thousands of studies conducted over the last fifty years that support my position about the effects of leadership on organizations. These studies show that subordinates prefer transformational leadership styles; leaders who inspire, engage, and consider their employees.

Study after study revealed that to achieve this leadership style, the leader needs to have practical experience. This can be developed in other leaders but it cannot be trained into them. These studies compared other leadership styles, and the results of these studies were depicted as "emotional" and "metamorphic" with respect to the transformational leaders, as "exchange nature" with respect to the transactional leaders, and through the lens of "permissiveness" with a laissez-faire leader. This is shown graphically in Figure 6.1.

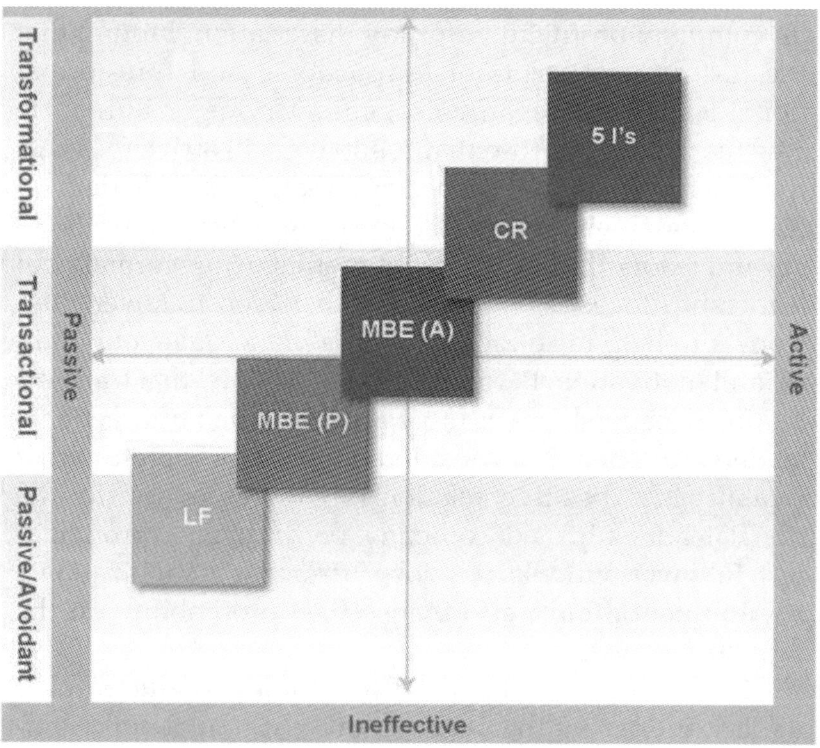

Figure 6.1 Full Range Leadership Model.

Transactional Leaders

Transactional leaders are more worried about maintaining the normal flow of operations. Transactional leadership can be labeled as "keeping the ship afloat." Transactional leaders use disciplinary power and an array of incentives to motivate followers to execute at their best. The term "transactional" refers to the truth that these types of leaders mainly motivate their followers by swapping rewards for performance.

A transactional leader does not look forward in strategically steering an organization to a place of market leadership; instead, these managers are only worried about making sure everything flows effortlessly today.

Transactional leaders are characterized by one or a combination of two styles. Management by exception is where the leader sets out rules and regulations and Key Performance Indicators (numbers); they manage by waiting for someone to step outside of these parameters and then taking action. This is sometimes referred to as the "gotcha" leader. The second style is the contingent reward. This is where the leader manages by offering a reward for certain performance. This is sometimes called "the carrot and stick" approach.

Transformational Leaders

In transformational leadership, leaders will elevate their power and influence using referent power (charisma) and visionary approaches (inspiration). The transformational leader is charismatic; the leader has a God-like quality that creates influence and referent power. Inspiration is another aspect of transformational leadership in which the leader possesses the ability to engage and emotionally connect and communicate an idealistic future state. This type of leader emits power and influences workers through visionary methods. Individual consideration is described by a leader's ability to gain power through mentorship and using a developmental approach with subordinates. Intellectual stimulation is derived from giving subordinates the power to solve old problems with new ideas. Employees are encouraged to question everything, including their leaders when appropriate. A subset of transformational leaders may be perceived as ideological symbols, representing something much larger than themselves to their subordinates. Leaders who are perceived as ideological symbols are less likely to be influenced by subordinates. From a role-modeling viewpoint, there may be moderate influencing from individual consideration and cognitive motivation prompted by intellectual stimulation.

Johannes Rank and collaborators examined variables that moderated the relationship between both transformational and transactional leadership behaviors and subordinate behavior in relation to their performance. Transformational leadership was a strong positive predictor for subordinate self-esteem, whereas subordinates had low company self-esteem. Numerous studies have shown positive correlation between individual productivity, unit effectiveness, and transformational leadership. In their study, it was hypothesized that a positive relationship exists between intellectual stimulation and inspirational motivation, and by extension, transformational leadership is important to motivating innovative activity in employees.

Assessment

Transformational leadership was assessed by employees, who rated their supervisor's behaviors using a 20-item scale. This scale included inspirational motivation, individualized consideration, idealized influence, and intellectual stimulation. Sample statements related to supervisor behaviors included the following:

- My supervisor talks optimistically about the future (inspirational motivation).
- My supervisor gets me to look at problems from many different angles (intellectual stimulation).

Transactional leadership was assessed by employee rating on a 5-point scale with 0 (never) to 4 (frequent). Sample questions included:

- My supervisor keeps track of all mistakes.
- I have trouble changing my behavior to suit different people and different situations in this organization.

The research focused on the organizational level of management innovation and the importance of leadership behavior as a precursor to organizational success, and proposed that management innovation may be a significant source of competitive success. Top Management Teams have the greatest influence on management innovation. Ignacio G. Vaccaro and associates examined transactional and transformational leadership styles with a focus on leadership behavior. The research concluded that company size can influence the impact of leadership and that both transformational and transactional leadership styles contribute to managerial innovation. Smaller companies tend to benefit from transactional leadership when innovating, especially when the company is less complex in structure;

larger companies, on the other hand, rely more on transformational leaders to counteract their complexity and facilitate management innovation.

Transformational leadership focuses on helping followers to identify with a purpose and goal. It stimulates the employees to achieve the company's goals by tempting their need for self-actualization. Transactional leaders participate in transactions as bargaining agents to achieve their wants and motivate their employees. They work to gain compliance from their employees by targeting their own interests and concurring with the conditions and rewards that follow the completion of requirements. Ultimately, the research of Vaccaro and collaborators supports the Hambrick and Mason position that leaders are internal change agents who implement policy, structure, and processes, and furthermore that organizational size is an important element of effective leadership, and particularly helps to define what type of leader may be effective in each environment.

Adaptive Leadership

Transformational leadership is at the core of adaptive leadership. The focus of adaptive leadership training is exemplified by high levels of moral and ethical conduct. Leaders demonstrating these values will gain the confidence of their followers, which will compel self-sacrifice for the larger good of their companies. These leaders are tasked with focusing on their own development and that of their followers in order to meet the challenges confronting their teams. Leadership, commitment, cohesion, and morale are critical components in unit performance. While transactional contingent reward leadership establishes the foundation for connections between followers and leaders in that it sets expectations, provides rewards and recognition, and clarifies responsibilities for accomplishing anticipated performance, Julian Barling and associates stated that transformational leadership boosts the development of devotees, daring them to think creatively and inspiring them to achieve beyond what they believe is possible. This is accomplished while reminding them to keep their high moral standards and values as guides to their performance.

A transformational leader reaches beyond managing day-to-day operations and fashions strategies for taking his organization or work team to the next level of execution and success. Transformational leadership styles focus on team-building, motivation, and collaboration with followers at various levels of an organization to realize change for the better. Transformational leaders set goals and incentives to energize their followers to higher performance levels while providing chances for personal and professional development for each follower.

One or more of four combinations of style characterize transformational leaders. Idealized influence is where the leader influences performance through a "charismatic" approach. This is sometimes called "walking the walk." The second style is intellectual stimulation. This leader influences performance by asking the followers to assist in crafting the design. The power of utilizing all the available resources is limitless, and maximizing this resource eliminates one of the seven wastes identified in lean (waste of intellect). The third style is inspirational motivation. In this style the leader inspires his followers to perform by showing them the way. It is sometime called "coaching" or "mentoring." The final style is individual consideration.

This leader takes an interest in followers as individuals and understands their wants and needs and can align them with the goals of the organization.

What makes a leader transactional or transformational? To answer this, you need to understand where leadership is learned. Our first source is our parents and family members. Forty percent of our leadership style comes from this source. The next sources are our teachers and coaches, then our community, business, and political leaders. Finally, there are professional athletes, entertainers, and religious influences. If you examine these individuals, you can better understand how you got where you are.

My Leadership Style

How did I make the leap from transactional to transformational? I credit the collective influences of several leaders. The first were my parents. They laid the foundation that I have built my leadership style on. When I left home and joined the Navy, I was lucky enough to have several good leaders to emulate. I find myself frequently using many of these early approaches in my daily tasks. After I left the Navy I found myself under the tutelage of a few more great leaders. They are still the folks I aspire to be in my technical and leadership approach. They never shut me down and always encouraged me to use my intellect. A lot of my professional confidence comes from their leadership and how they developed me into the leader I am today. When it comes to the type of leader I want to be in my chosen profession, I find myself looking to my last true general manager. I worked for him during my tenure in the mines. He always walked the walk. His idealized influence over me still burns bright. I find myself reflecting on what he would do in each situation. He worked his way up the professional ladder, holding every position from entry level to his current title of Chief Operating Officer for a leading organization in his industry. Along the way, he has developed a list of senior leaders who credit him for their leadership styles and successes. For every one of these great examples, I have had an equal number of bosses who I have worked for but never emulated or followed. The difference between success and failure is your ability to follow great leaders and know which bosses to use as bad examples.

The informative years for all of us are from birth to five years old. Think about how you are taught during these years. You have a coach/mentor by your side guiding you. If you do something well, you are praised and even possibly instructed as to why you did well. If you attempt to do something harmful, you are stopped and instructed in corrective measure to prevent harm. Sometimes you can fail, and then your coach helps you understand why you failed and how to adjust to correct the approach. This coaching continues throughout your first five years.

What changes when you turn five? You are sent off to school. Here the teaching changes. In your academic environment, you are rewarded for good behavior and disciplined for bad. This is repeated by your leaders and others in leadership positions serving as examples for us. No wonder most of us end up leading transactionally; it is all we know.

CHAPTER 7

Things Leaders Should Stop/Start Doing Immediately

With the increasing pressure to do more with less, we are bombarded by the need to change the way we think. To lead organizational change, you need to figure out what you should stop doing. The difficulty is that this process is tougher than it appears. Many of us must manage both up and down the chain of command. You may find yourself in an organization that is not ready for a communication approach that is vastly different from one that anyone has seen before. With the preponderance of senior leaders lacking leadership skills, you may need to educate the leaders of your organization before employing the recommendations that follow.

WHAT TO STOP DOING

Five things I believe all leaders should stop doing in order to become more effective:

Best Practices

There exists no better example of herd mentality than that of best practices. I have always thought that a so-called best practice stops the minute it is branded as such. Best practices defend the status quo and limit innovation by ensuring people/processes follow the same practices. You cannot distinguish by accepting likeness. The idea of best practices is nothing more than mixing norms at your own peril. Smart leaders innovate past best practices, always in search of next practices. If your choice to do something is made because others are doing it the same way, you are doing nothing more than conceding advantage and opportunity to those competitors more creative than you. Lean manufacturing, and most other recent initiatives that are embraced by senior leaders, is focused on driving the organization towards best practices. In our rush to shortcut hard work, we have conceded competitive advantage. To regain this advantage, we need to realize this and stop doing it. Don't copy, create.

Cost Cutting

It is just impractical to beat your opposition to the future by spending less than they do; you get there first by investing more smartly than they do. Companies who surpass their competition focus less on risk and more on opportunity. They are less concerned with controlling expenditures and more worried about finding new ways to create greater return on investment. I have often promoted the idea that the duty of leaders is not to leverage their people but to establish more leverage *for* their people. Stop expecting your people to do more with less and find ways to provide them with a resource advantage.

There is a big difference between cost cutting haphazardly and reducing the budget by improving reliability. To help you make budget decissions based on actionable information, I have included a budgeting template on *www.deathofreliability.com* to give you the ability to track your performance to your budget. I have also included a check book template to accurately control your costs. Stop levying hiring freezes and begin an unyielding pursuit of creating a talent advantage. Leaders who complain about a lack of resources are doing nothing more than demonstrating their lack of resourcefulness.

Political Correctness

The fact is that politically correct thinking is most often deceitful, if not altogether intellectually dishonest. Politically correct thinking replaces uniqueness and authentic opinions with socially acceptable rhetoric and diluted behavioral tendencies. I miss the days when most conversations consisted of highly-charged and stimulating discourse where people were urged to openly share their true thoughts and opinions. The irony of politically correct thinking is that a society empty of individual thought creates the reverse of diversity. Politically correct thinking results in a programmed group of sheep who completely lack diversity because of a gentrification of thoughts and actions. The sinister secret behind politically correct thinking is that it gradually clouds your senses, and sterilizes your inherent capability to be discerning. If you are like me, you don't want your team to say what they think you want to hear, or what they believe they should say, but you do want them to say what they are thinking. How many meetings have you attended where everyone sat around the table like a bunch of deer in the headlights trying to figure out how to wiggle around an issue rather than address it head on? It is this type of issue that taints our culture, suppresses innovation, weakens our productivity, and condemns those who accept politically correct thinking to a life of mediocrity.

Glorifying the Few

Leadership is not a position or a title. It is not a job reserved only for a precious few presiding over the masses. Here is something to keep in mind: if you tell people enough times or loud enough that they are not leaders, you should not be amazed when they begin to believe you. Your job is not to repress people from leadership, but to establish leadership ubiquity. The most triumphant organizations are those where all team members view themselves as leaders. Leadership that cannot be transferrable, scalable, repeatable, and sustainable isn't leadership at all. Build your organization on a foundation that builds leadership into all team members, regardless of where they are on the organizational chart.

Unwillingness to Change

Look at any study on the speed of change, and you will find you are living in an unprecedented time. The rate of change is clearly overtaking most leaders' ability to learn and unlearn. Most leaders fight to remain current, not to mention find a way to move ahead of the curve. Here's the thing: if leaders are living in the past, their organizations will be required to travel an extremely rough road to the future.

The solution to the leadership training problem is to scrap it for development. Don't train leaders—mentor them, coach them, and develop them—but please do not train them. Where training attempts to homogenize by blending to a norm and adapting to the status quo, development attempts to call out the unique and discriminate by shattering the status quo. Training is something leaders dread and will avoid, while they will embrace development. Development is nuanced, contextual, collaborative, fluid, and primarily actionable.

The reason it is important to stop doing things that do not add value is to free up the time you will need to start doing things the right way. People are always telling me, "This is great but I do not have time to do what I am doing now." I have found myself in this same place but I eventually realized that about half of what we do each day does not add value. Taking a critical look at your daily activities will open your eyes. How many meetings do you attend that have other members of your team in them? Why? How many meetings do you attend and say nothing? This goes on and on. When I say this I always hear, "Not me. Everything I do adds value." Sorry, you are lying to yourself, because that is not possible. The only way to make a change happen is to do something. To do something, you need time. To get the time, you must stop doing things that do not add value. As a leader, you need to understand what adds value. The key to adding value is to take ownership and not use victim statements. Next, I want to explain what I mean when I say that you should take ownership and not be a victim.

START TAKING OWNERSHIP

The ability to lead is all about taking ownership. Throughout your day, you must make many decisions. To accomplish this, you must make a choice. This applies to almost every moment. You stand at either a conscious or unconscious nexus of choice. At this moment, you have to choose to respond as either a victim or an owner. The criticality of this decision cannot be overemphasized. Even when they seem unimportant, these choices represent pivotal moments in your life. In these moments, the victim or owner decision has consequences, and these consequences drive results. With each owner choice, you take charge of the issue, costs, and outcomes by deciding to act rather than be influenced. A good friend of mine, Dr. Dennis Deaton, is the author of the book *Ownership Spirit: the One Grand Key that Changes Everything Else*. When I first read his book, it changed my life. Dennis says, "The consummate truth of life is that we alter our destiny by altering our thoughts." We have power over our thoughts, and while you may not be able to control what happens to you, you can control how you respond to it. I am reminded of this day in and day out.

Reliability leadership is all about making decisions. The only way to accomplish this is to have the experience to draw upon to make the decision. The only way to get this experience is to have hands-on reliability work history. There is no college degree that teaches you what you need to make the necessary decisions every day as a reliability leader. As I have noted over and over throughout this book, there are no shortcuts to leadership in reliability. There are no substitutes for hard work and the time it takes to become a skilled tradesperson. Organizations are setting the people they are hiring to fill these key positions up for failure if they are not qualified as defined in Chapter 5. To have the confidence to take ownership of the situations that they will be presented with, they need the wealth of knowledge that only comes from hands-on experience. A title does not make you a leader, having followers does, and the only way to get a reliability team to follow you is to earn their respect. To do this, you will have had to have walked in their footsteps and had the internal fortitude to take ownership of the fact that experience is the only path to reliability leadership.

OWNER OR VICTIM ARE NOT CHARACTERISTICS

Owner or victim are not characteristics but represent contrasting mental approaches. They are different ways to look at the world that deliver dramatically different results. When you are a victim, you relinquish your power and are ineffective. As an owner, you retain your power and release more power. Think of all the times you have been confronted by the issue of having insufficient time and money to accomplish a task. How did you respond? Was your response *"How do they expect us to do this,"* or, was it *"Here is what we can do given these constraints?"* The difference between victim and owner thinking is that simple.

The biggest part of this battle for me has been realizing when I am victim thinking. It is not possible to be an owner all the time. No matter how hard I have tried, I have not been able to keep the victim inside me from slipping out from time to time. The key to managing the times you are the victim is to realize when it is happening and make the shift to being an owner. Now does that mean you need to wait for things to happen to you? Absolutely not. Here is where experience can help you prevent situations where you find yourself making a victim or owner decision. This is another example of being an owner. It is not all about how you react, but how you approach every moment of your life. Having the experience to get ahead of situations is another aspect of taking ownership. Here again is where organizations are hiring into leadership positions people without the prerequisite experience and setting them up for failure. While owner and victim are not identities, they complement your identity either for better or worse.

A victim looks for a scapegoat. There are a lot of excuses, from something nearby all the way to God. Victims are pessimists who look at the world with a "glass half empty" mentality. They squander their lives, viewing it as one big plot or a string of hindrances. When leaders act as victims they feel that all situations are overwhelming and that they cannot influence anything. Victims reject accountability for the consequences and do not take responsibility for anything. You hear victims say things like *"They made me do it," "I had to do it,"* or *"It is your fault."* When victim thinking, you fail to work on problems, goals, or challenges. Victims are indignant, astonished, and intimidated. When victim thinking, you do not make a difference or see opportunity.

In reliability leadership, there is little room for victim thinking. When someone is hired for a position that they do not possess the knowledge or experience to fulfill, they will quickly find themselves in the victim-think mentality. Because they lack the breadth of experience to draw upon when confronted with the day-to-day operations of reliability efforts, they quickly look for scapegoats. I have heard many such persons say, *"I need to ask my manager,"* or *"How am I supposed to know that?"* Qualified reliability professionals do not need to ask their managers because they are the subject-matter experts themselves. The answer to *"how am I supposed to know that?"* is experience teaches you. With qualifications comes more owner speak and thinking, so organizations looking for their leaders to take ownership need to hire qualified people for these positions.

HOW TO THINK LIKE AN OWNER

Owners look for what to do. They know that the right thing is usually not the easy thing to do and they don't look for it to be. When they are faced with an obstacle, they figure it out and overcome it. Owners concentrate on values that trigger success and learn from previous lessons to build their body of knowledge. The enjoyment of overcoming resistance is exactly equal to it. Referring back to my Navy SEAL example in Chapter 5, SEALS are the epitome of ownership spirit.

Making the decision to be an owner comes with experience and the knowledge that you are qualified for the tasks at hand. In the world today, organizations are less worried about having the most qualified personnel in their positions and more worried about the bottom line. I have personally seen this in most of the organizations I have worked for and with. Few of the organization's leaders have earned their way to the positions they hold. Most are examples of the "Peter Principle;" that is, they have been promoted to their highest level of incompetence. All levels of these organizations suffer from having leaders who do not have the knowledge or experience to execute their jobs effectively. Because of this inadequate leadership, these organizations lean towards victim thinking. Owner thinking eludes them because they lack the qualifications for the positions they hold.

You share the process of victim or owner thinking. Nobody is one or the other but a combination of both. You move through victim thinking so easily you don't even know you are doing it. It is important to understand that no one has a set personality or fixed characteristics. Once you understand that what you are doing works and is progressing, you will be prepared to smash through all obstacles. There is no limit to the chances to release the power of victim/owner choice. Developing ownership muscle gives you the capability to act calmly in conflict or the noise around you. It allows you to think clearly. You have the option to respond as an owner or react as a victim. Real reliability professionals respond as owners.

CHAPTER

8

Managing Consultants

I have often marveled at an organization's willingness to turn over its reliability efforts to outside consultants. Imagine you are hiring someone to lead your reliability department. There are two final candidates for the position.

Candidate number one has 14 years of experience. He has never worked in an industrial reliability environment. He holds no journeyman skills or licenses and has never held a position as a planner, supervisor, or manager. He has only theoretical experience in any industrial environment. His work experience has been as a consultant. His education is a bachelor's degree in computer science.

Candidate number two has 20 years of hands-on, in-the-field reliability experience. She holds a journeyman license and has extensive journeyman level skills. She has been a superior performer in the reliability industry for her 20 years. She has worked as an apprentice, journeyman, planner, front line supervisor, senior supervisor, and manager of maintenance and reliability. She holds a bachelor's degree as well as a MBA. She has led reliability transformation for CAT, GE, and Ford, to name a few. She is a certified maintenance and reliability professional and a licensed project management professional with multi-million-dollar project leadership and success. She has an ANSI certification in machinery lubrication. She is a sought-after speaker on maintenance and reliability and a subject matter expert in maintenance and reliability.

As a senior leader looking for the best person to lead your organization to achieve reliability, who would you hire? An overwhelming number of organizations are hiring candidate number one, the consultant, over their internal resource. They will pay hundreds of thousands of dollars for someone who has no chance of achieving the desired results. Organizations will scrap months of work in reliability in favor of someone who has never spent one hour on their site and did not do any due diligence to understand what needed to be done, but who assumed his company's antiquated model was the right thing for your site. This failed approach is repeated across all industries every day, because senior leaders lack the knowledge and experience to make the right decision. What follows is an outline of the pitfalls encountered when managing consultants and how to avoid them.

THE WRONG APPROACH

Because of their shortsighted decision to shortcut hard work, organizations are undertaking reliability projects without true reliability professionals leading them. This will result in the project falling short of the desired expectations. To combat this shortfall, organizations turn to consultants to lead these efforts. The problem with this approach is that they are no better off in following this path. Webster's Dictionary defines a *consultant* as an "experienced professional who provides expert knowledge (often packaged under a catchy name) for a fee. He or she works in an advisory capacity only and is usually not accountable for the outcome of a consulting exercise." When an organization hires a consultant, it should be based on merit, and the organization must manage the consultants carefully. The internal staff of the organization must be the focus of the transformation, and the consultant solely a resource to ensure that the goals set for the project are met or exceeded. The only real way to manage a consultant is to hire a qualified reliability professional. The bright side of this is if you hire a qualified reliability professional you will not need the consultant. This is another significant cost savings and a source of funds to invest in your improvements.

UNDER PROMISE AND OVER DELIVER

To further operational objectives, senior managers will employ consultants. Consulting firms rarely deliver the company's expectations expressed at the initiation of the project. The main cause of this is that the organization fails to manage the project. Consultants sell a model structured to prevent the realization of the project objectives. Their model delivers on their strengths regardless of whether they addressed the company's needs. I am not surprised that companies fail to use consultants correctly to get the best bang for their buck. Companies never measure the consultant's performance against expectations, but consultants are masterful at lowering expectations to close the gap between what the company expects and what the consultant can deliver. The consultant's goal is to "under promise and over deliver," but they usually deliver the opposite.

If companies base employing consultants on their needs, they end up staffing them in roles as functional experts, which prevents sustainable results. When they employ them to help achieve performance levels that could not be achieved without them, the company is actively involved, and sustainable results are achieved. This sounds simple but is rarely achieved. If you are hiring consultants because you do not know what to do, you should stop. Consultants should only be hired if you have qualified internal resources to manage them.

FAILED POLICY

Without a true reliability professional guiding the organization and identifying the roles where the consultant adds the most value and sourcing resources for the other roles, the project is doomed to fail. Clearly, the key ingredient when engaging a consultant is having a skilled reliability professional leading the project to avoid diluting the project objectives with unskilled and inexperienced consulting firm resources. Without the right internal leader, the consultants will quickly dismiss the capabilities of the company's personnel, when in fact the company possesses the right folks to manage and move the project forward and facilitate the buy-in to the process. Companies fail to understand that leadership is not demonstrated by hiring a consulting firm but by applying their own considerable knowledge and resources as the best approach to embedding expertise in the project while controlling costs and effectively managing their available resources.

A company needs to have a leader with the expertise necessary to evaluate the consultant's proposal and its ability to deliver on the project's objectives. Consultants carefully craft their credentials and resumes to give the appearance that they are the perfect fit for the role. Taking refuge in a favorite or familiar brand rather than investing the time to verify that their team can deliver is another failure commonly found when organizations do not have the right person making these important decisions. Even when members of an organization have previous experience with a consultant, past performance does not guarantee success. Each organization has unique challenges and falling for a consultant's model one-size-fits-all approach is yet another failure of most organizations. These are the purchasing behaviors that consultant firms encourage in their clients. Shouldn't a company be able to purchase what they need, instead of what a consulting firm is selling?

FAILURE TO CONTROL

If you do not control consultants, they will control you. Irrespective of the extent of value-added expertise applied, the value realized by the organization will be determined by the level of control the company can exert over the consultant from the first contact to final invoice. Less than four percent of executives, managers, and employees have the education, training, or experience necessary to guide them in managing consultants. The largest number of executives, managers, and employees spend most of their time on the day-to-day operation of the business. They have limited participation in projects where consultants are used and even when they do it is infrequent. Even those assigned full time to projects do not have the necessary expertise.

Most of these individuals come from operational roles and have never worked in reliability. Companies persistently allow consultants, who manage their clients every day, to be opposing internal personnel, who do not have the experience to manage consultants effectively. Companies need to promote leaders into the positions they are best suited for and not simply promote them because they need to fill a position. Some companies assume that the responsibility of managing consultants rests with procurement. However, consultant services are not commodities and need to be paired with an internal resource capable of managing the resource. The proper pairing will determine the success of the project. Long-tenured employees most often are not the correct resources to lead or manage a reliability project since the need for the project is due to the lack of capability within the organization. Organizations should look to internal resources with the qualifications to manage the project regardless of company position.

WHO IS LEADING?

Too often, consultants lead and control the relationship. If the client does not control it, then it is highly unlikely that the project will meet or exceed the original expectations. An honest assessment is required before beginning a project. Organizations rarely have an independent review of a consultant project either before or after completion. They rarely analyze how the project was executed, assess the outcome, and tally up the total costs compared to the original business case. Most company boards fail to ensure that there are appropriate policies in place to govern the selection of the consultant despite the impact they are expected to make on strategy, cost reduction, or growth. Thus, consulting projects usually fail to achieve the desired outcomes more often than the company realizes. Consultants do not highlight their own underperformance and company executives, managers, and employees similarly want to be connected with the illusion of project accomplishment. Companies routinely report projects to be more successful than they are. Companies would be surprised at how often consultants fail to deliver, especially if they analyze the various projects that have been viewed internally as a success.

To succeed, an organization needs to surmount the pressures driven by the typical consulting firm's model. For instance, executives must be aware that consulting firms will assign experts depending on the jargon of the organization in question to win the project. Once they have won the project bid, they will leverage those scarce experts across other clients. This leaves a full complement of less qualified consultant resources. Heavy advertising, carefully crafted brands, volumes of published leadership material (often produced by external agencies unrelated to the consulting firm), and long-time executive relationships have all diverted clients' attention away from the abilities of individual consultants. Senior executives need to be careful not to sign substantial contracts with consulting firms that are based on concepts and intellectually elegant value schemes without fully exploring the capability of the team to execute and realize the desired business outcomes. Teams produce projects, not brands.

Senior leaders should be aware that most consulting firms have a core capability around which they are trying to sell. But as they actively increase their relationships with the client company and try to cross-sell, up-sell and follow-on sell, consulting firms, fueled by aggressive revenue-growth targets, will often end up offering services in areas further and further removed from their core capability.

The abilities of global consulting firms can also vary considerably by geographic region. For example, almost none of the so-called "global consulting firms" have the extent and depth of sustained skills across the diverse markets in Asia that they have in Western markets.

Selecting consultants centered on brand and relationship, without studying the specific requirements for the geography or service area, is apt to result in underachievement and perhaps overpayment. Leadership is required to prevent making what are falsely presumed to be safe selections simply based on brand or relationships. Consultants should be engaged based on which firm has the most qualified resources that are fully available for the task at hand.

CONSULTANT RESULTS

Organizations that run headlong into hiring a consultant without having someone within their company with the expertise to manage the relationship fall victim to biases introduced by the consulting firm's model. This, coupled with a lack of management skill on the client's part will result in:

- A failure when junior consultants are hired in roles that could be obtained more cost-effectively in another place. Compounded by an absence of supervision and management by the client, this results in a complete project failure or a result that considerably misses expectations.

- Overpayment for the project objectives that were only mostly achieved. Costly consulting resources were used in augmented roles, so the project costs more than it should have.
- Underachievement results are lower than what could have been realized with the assigned resources, often because the scope was not managed and the client did not actively manage the consultant's activities.

Only when leaders engage consultants based on their merit, exercise influence on their consultants (beginning with control of how they use them), and keep their team members engaged as integral parts of the project can an organization expect consultants to meet their expectations. Your team and the consulting staff work together as an integrated team, with the consultant focused on facilitating, advising, knowledge-transfer roles and tasks that require expertise that your team does not have. At the same time, your team actively controls the resources, the relationship, the deliverables, and the project.

Organizations will improve their history of accomplishment in receiving full value from consultants if they focus on increasing their level of control over their consultants and their project, and if they manage how the projects are staffed and which roles are filled by consultants. This requires leadership, participation, and discipline on the part of management. Effective consulting requires effective leadership.

Be Wary

Be wary of consultants who tell you that you will save more than you spend. This is another way they manipulate you and your expectations, and they will never deliver on this statement. Most organizations fail to budget specifically for improvement efforts. They spend important reliability dollars chasing exaggerated promises that have no substance. This approach is repeated over and over again, and it produces little or no results. Why do companies think they are successful? It is because the consultants are masters of manipulation and lowering expectations and they are aided by internal resources charged with making these initiatives succeed. The internal resources are driven by lack of skill, knowledge, or qualifications. They fear the loss of their jobs and are reaching out to anything that appears to offer a life preserver. Sadly, consultants are no more a life preserver than a block of concrete. Between the two they lie about results and mask failure with distorted outcomes. I know some of you are thinking I am wrong, and I will bet this is because you are either a consultant or one of these internal resources. Spending the limited reliability dollars on consultants and their fantasy programs is a complete waste of reliability money. It is a poor decision on all fronts. If an organization is foolish enough to fall for one of these trumped-up efforts, then they believe it is the right thing to do. Fund the effort outside of reliability so when it fails to achieve the promised results, at least your reliability efforts are not shorted. If you find yourself being asked to hire a consultant, you should know you are being set up to fail. If it is a subordinate asking you, this person is lazy or unqualified. If it is a superior, the same words apply.

The goal of all consultants is to reduce your expectations in order to close the gap between what you need and what they offer. It is my belief that organizations should stop looking for shortcuts that do not deliver on their needs and start looking for solutions to their issues. Shouldn't companies be able to buy what they need, rather than what a consulting firm is selling? The way to accomplish this is to look for solutions. The difference here is that a solution is something that is done to deal with and end a problem. It is something that solves a problem. This is in stark contrast to the trendy hiring of consultants to fix your problems. I suggest that you take ownership and start solving your own problems.

Having said all of this, I want to clarify that I am not attacking the individuals that are employed by the consulting firms. These individuals are mostly good people trying to perform to their company's expectations. I have an extensive network of resources that would be classified as consulting firms, and I trust each and every one of them. Like I have said repeatedly in this chapter and throughout my book, the key to successful use of a consulting firm is having the right internal resource managing these efforts. I use consultants in efforts that are resource intensive or where I need additional expertise. This is a scalpel approach, not a broad meat clever. If you believe you need to allow a consulting firm to reinvent, develop, or run your reliability efforts, you should save your money and invest in hiring a qualified reliability professional to lead your team. This is a sustainable approach and hiring a consultant is not.

CHAPTER 9

Reactive Maintenance Strategies

RUN-TO-FAILURE MAINTENANCE APPROACH

A run-to-failure maintenance plan is one that essentially operates on the run-to-failure strategy. This is a very hands-off approach to machine maintenance and while it keeps routine maintenance costs low, such a program is the costliest overall to an organization. This is the easiest maintenance approach to install, but it will cost the company the most. This is the most recognized of the reactive maintenance approaches. Because of the prevalence of this strategy, I will spend more time discussing it than other reactive approaches. The driving factor in an organization's decision to employ a run-to-failure maintenance strategy is the fact that they lack a true reliability leader. The need to follow this approach is rooted in the inexperience of the organization's assigned maintenance or reliability manager. In the haste to shortcut the hard work necessary to achieve reliability, organizations appoint unqualified engineers or other managers to this key position and pat themselves on the back for saving the time and money while spending thousands more replacing failed or failing equipment.

The Department of Energy's Office of *Energy Efficiency and Renewable Energy* (EERE) notes that the two major advantages of run-to-failure maintenance are reduced maintenance costs and the need for less staff members. However, these are all of the advantages and they are outweighed a hundred-fold.

Run-to-failure maintenance is the process of reacting to failed, ineffective or damaged equipment and repairing or replacing it so the intended function can be achieved. Essentially, run-to-failure maintenance ignores any proactive measures and simply deals with a problem or issue when it is reported.

Studies as recent as 2000 indicate that run-to-failure maintenance is still predominantly the mode of maintenance in the United States. One study breaks down the average maintenance program as 55% run-to-failure, 31% preventive, 12% predictive, and 2% proactive. This is displayed graphically in Figure 9.1. This study emphasizes the enormous opportunity awaiting any organization looking for a competitive advantage and the imperative need to hire or promote a true reliability professional to lead the transformation to proactive reliability.

There are a lot of issues with this approach to maintenance. The following section outlines the advantages and disadvantages of run-to-failure maintenance.

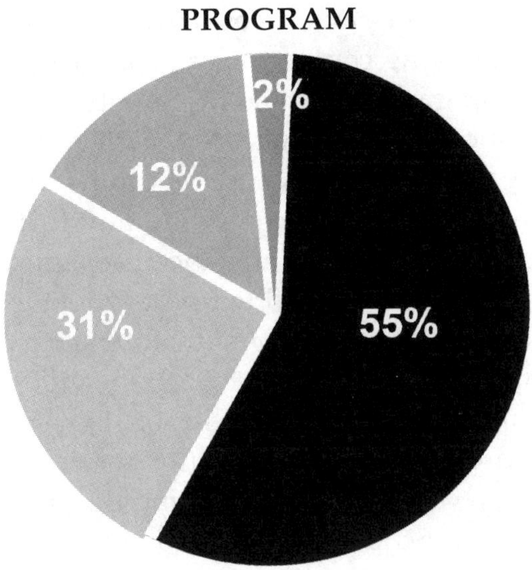

Figure 9.1 Maintenance Strategies Breakdown.

ADVANTAGES

The first of the advantages in running a run-to-failure maintenance system is the lack of initial cost involved. When only reacting to a problem, if the point of contact or the relevant contractors are known, there is very little to do before a problem occurs. Therefore, investment in systems, procedures, pre-planning and organization time is greatly reduced, along with the startup cost required to begin the operation. These cost savings are eclipsed by the significant costs associated with replacing failed equipment.

The limited personnel requirements, due to the reduced planning, management and organization time involved with run-to-failure maintenance, requires less staff to manage the facility. This can offer senior leaders the opportunity to save on payroll costs. Again, this is dwarfed by the cost of having contractors respond last minute and is often accompanied by extended downtime.

Maintenance costs are reduced with a run-to-failure maintenance approach, because equipment is used as much as possible without expending any further money than its purchase price. Maintenance is evaded, servicing and checks are not done, and temporary repair and quick fixes are made to reduce ongoing maintenance costs.

Some operations will favor dealing with run-to-failure maintenance for the possibility of higher profit margins. There are contractors that offer services and prefer an 'emergency call out' and overtime due to the chance to charge more for the repair, especially when the client is desperate for the repair to be completed and is not in the position of being able to bargain or get a second opinion.

Disadvantages

Unpredictability is one of the main disadvantages of run-to-failure maintenance. This lack of knowing may result in either increased labor or materials being unavailable instantaneously and therefore delaying the time needed for a repair, increasing equipment downtime. Furthermore, off-hours or emergency visits by contractors to fix an issue can result in needlessly expensive charges that may have been avoided. This aspect of unpredictability can also influence budgets and cost planning as there is no maintenance plan and anticipated cost in place.

Equipment is not maximized in run-to-failure maintenance. It doesn't protect or look after equipment and therefore reduces the life cycle of the unit. Rather than maintain the equipment and ensure it is running in optimal condition, a run-to-failure maintenance approach does the bare minimum to keep equipment operational. The negative effect of this is that the equipment won't fulfill its potential or return on investment.

Indirect costs are found with run-to-failure maintenance with unreliable equipment and equipment downtime causing negative effects on reputations, the ability to run a business efficiently, productively, and safely. Furthermore, run-to-failure maintenance can be an unproductive use of employee time, both for the operators of the equipment and the management personnel who must work hard to arrange for repairs to be made.

What Are the Alternatives?

An alternative approach to run-to-failure maintenance is preventive maintenance, which takes a different approach, with the emphasis being on taking steps and measures to avoid any breakdowns or problems occurring with equipment. By servicing, and making regular checks and replacing parts when recommended, a preventive maintenance approach limits unexpected problems and aims to reduce long-term expenses through avoiding unanticipated issues.

However, it is worth noting that preventive maintenance cannot be solely relied on as there are always going to be unforeseen breakdowns or issues. For this reason, every reliability management operation will need an essence of run-to-failure maintenance.

Costs will happen in several areas using this approach. When a machine fails without warning, it creates downtime within the plant, which can get very expensive. This also increases labor costs to get the equipment back up and running, especially if a worker must do overtime.

The repair and replacement of equipment also becomes an issue, and in some cases, back-up equipment is necessary to keep the plant running. Ultimately, this is a wasteful use of staff as it requires workers to stop what they are doing to address the problem.

The driving force behind the predominance of run-to-failure maintenance strategies is the lack of leadership in most organizations today. The desire to shortcut the hard work and effort required to use reliability to a competitive advantage are viewed as too costly. Since no one in leadership possesses the knowledge or skill to lead the organization down this path, the decision is to "check the box" and give it lip service in a vain attempt to display the appearance of a proactive reliability strategy. This can have a short-term appearance of success because those viewing the results do not possess the experience of knowing what an honest, proactive strategy looks like. As I have repeatedly stated throughout this discussion, the lack of leadership is the root cause of an organization's decision to employ a run-to-failure maintenance strategy. I have used the decision tree in Figure 9.2 to help determine which reactive strategy is best, given the situation. The only way to overcome this failed strategy and loss of competitive advantage is to hire an experienced reliability professional to lead your efforts.

PREVENTIVE MAINTENANCE APPROACH

Preventive maintenance is guided by time, meter or event based activating. Maintenance that is done during PM's is pre-determined based on several factors including age,

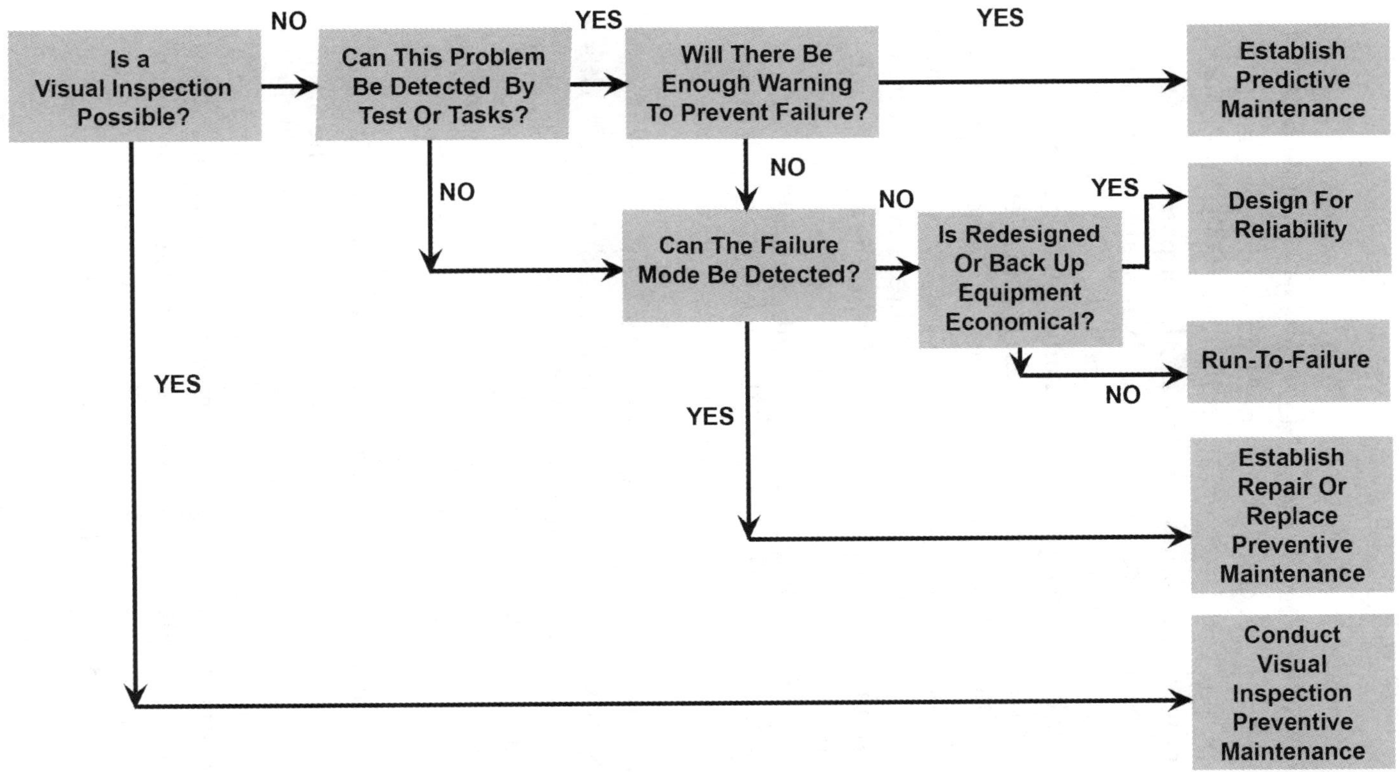

Figure 9.2 Reactive Maintenance Decision Tree.

manufacturer's recommendations, experience, etc. It is expected that a machine component will fail over a specific time period that is common for its type. Under a preventive maintenance approach, the pertinent parts will be replaced, removed, or rebuilt on or before the predictable failure point. *Mean-time-between-failure* (MTBF) statistics can help optimize a preventive maintenance management schedule to include inspections, repairs, and rebuilds. Like run-to-failure, this too is a reactive maintenance approach. Since 90% of failures are random in nature, this approach offers little assurance of reliability. There is no component of this strategy that proactively addresses equipment reliability.

The principal issue with a preventive maintenance approach is that the operation of the machine, how often it is used, as well as many other variables, will influence the operating life of the machine and its components. In most cases, parts are replaced before they need to be. Therefore, this approach results in unnecessary maintenance. This unnecessary maintenance costs the organization money, and if the technician causes collateral damage during the replacement that creates additional costs. Forbes magazine reported that "one out of every three dollars spent on preventive maintenance is wasted."

Preventive maintenance needs more ongoing effort, but when done properly, it can reduce costs in both the short and long term when compared to run-to-failure. While it does not eliminate the risk of machine failure, a plant has a better chance of catching and correcting issues before they fail using this strategy. This approach requires more labor to perform preventive tasks that may seem unnecessary, but as already stated, this extra effort pays off overall.

Preventive maintenance involves maintaining equipment by tending to its performance at specified time intervals, much in the same way you bring your car in for an oil change every 3,000 miles. This will make sure that certain machine elements, like lubrication, are always in good condition and not wearing down. Like all OEM recommended preventive maintenance efforts, changing lubricants on a time-based interval is needless. This is where a more proactive approach driven by oil sampling provides the most cost-effective solution.

The main advantage of a preventive maintenance plan is cost effectiveness when it is compared to the most expensive maintenance processes (run-to-failure). It will save energy because the machines will be operating more efficiently. Preventive maintenance will extend the life cycle of a machine and reduce having to purchase larger and more expensive replacement parts. Overall, this approach generates an estimated 12 to 18 % savings over run-to-failure maintenance strategies, according to the EERE study.

The Problem

The problem with employing a strictly preventive maintenance strategy is that it gives the organization the illusion that it is doing something proactive. There are significant differences between preventive and proactive maintenance strategies. Preventive maintenance replaces components before their failure at intervals. This is mostly a glorified run-to-failure approach. The only difference is that the component is removed before failure if the intervals are correct. This approach does not guarantee your equipment reaches its designed end of life, and it reduces this life in favor of early replacement. By employing a preventive approach, you make the decision to increase your costs by prematurely replacing components. *Hydraulics and Pneumatics Magazine* stated that "sixty percent of hydraulic pumps sent in for repair had nothing wrong with them."

The advantage of preventive over a purely run-to-failure strategy is the cost savings on collateral damage and the delays caused by the resources necessary (parts and contractors) to affect the failure that was not planned. It is these savings that make preventive maintenance 12-18% more cost-effective over run-to-failure.

PREDICTIVE MAINTENANCE APPROACH

Predictive maintenance is defined by monitoring the condition of equipment based on either its average or expected life statistics. Essentially, this methodology tries to predict the failure before it happens by directly monitoring the machine during normal operating conditions. Referring back to the car example, rather than replacing the oil every 3,000 miles, when using a predictive maintenance methodology an oil sample can be taken at predetermined intervals, and the oil is replaced when it decays beyond a certain point. This is called *condition-based maintenance*. In many cases, when predictive analysis notices an issue, the repair can be scheduled at a time that minimizes the impact on production. Like the previous two maintenance approaches, predictive maintenance is a reactive approach.

A variety of cost-effective tools can be used within a predictive maintenance program to analyze system performance and generate accurate factual data about critical systems so that maintenance activities can be scheduled the moment they are needed. Improvements in profitability, quality, and productivity can result when predictive maintenance is implemented on capital-intensive assets.

Predictive maintenance is an enhancement upon preventive because instead of replacing a component at a set interval, it provides condition-based information to push the replacement of the same component closer to its failure point. Predictive maintenance employs several tests to allow the planned replacement of the failing component before complete failure. Like the previous reactive maintenance strategies, predictive maintenance needs the component to begin to fail in order to track it.

EFFECTIVE MAINTENANCE MANAGEMENT

Many businesses are surprised to learn that considerable maintenance dollars can be saved by employing effective maintenance management approaches in the right places. The effect of random or inappropriate maintenance can undoubtedly be experienced in terms of the quality of products and services, operating costs and thus the bottom line, so it makes sense to select the appropriate policy for each asset. I have included the basics of managing maintenance work in Figure 9.3.

Predictive maintenance involves inspecting equipment with a variety of test equipment including infrared and ultrasound technology. A NASA report showed that this maintenance approach works to eliminate unexpected breakdowns and helps schedule maintenance down time that would otherwise be spent to perform more invasive maintenance approaches.

Predictive maintenance can be equated to a checkup at the doctor, using sophisticated technologies to gather information on the health of the equipment. This can help

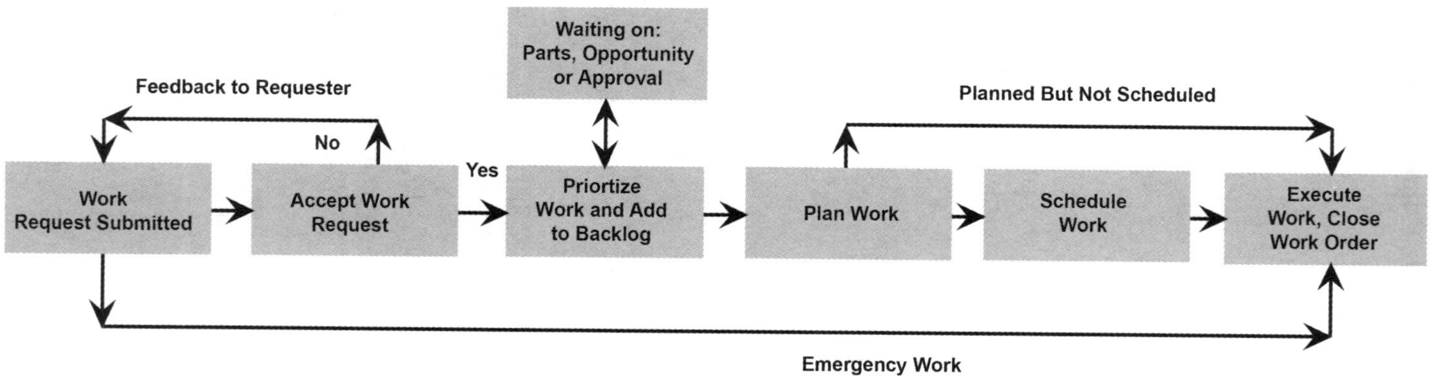

Figure 9.3 Work Flow Management.

indicate where possible wear and tear may occur and assist in identifying a failure before it happens. The NASA report showed that this maintenance approach alone can reduce maintenance program costs by 50%.

CHAPTER

10

Proactive Reliability Strategies

PROACTIVE RELIABILITY APPROACH

The three previous maintenance strategies are all reactive. The run-to-failure strategy is usually referred to as reactive. Classifying preventive and predictive maintenance as reactive creates a lot of discussions. I believe that there is a simple way to see how all three of these strategies are reactive. As I already said, run-to-failure is universally accepted as reactive.

Preventive and predictive maintenance require the component in question to be in a failure state. Preventive maintenance is a time-based strategy and these time-based intervals are based on failure time periods. Thus, while the preventive (time-based) replacement strategy has the appearance of being proactive, it is most definitely not. The time-based intervals are based on catastrophic failure and this is reactive. Before a component catastrophically fails it enters a failure mode and its performance is degraded. Preventive and predictive maintenance can save you 12 to 50% over run-to-failure.

This same analogy is true about predictive maintenance. Predictive maintenance uses advanced technology to monitor and track the degradation of components and recommends replacement before a catastrophic failure. Again, like preventive, it is reacting to the component's failure. While mislabeled by many unqualified posers in reliability leader positions, it is reactive. You can achieve a 50% reduction in costs over run-to-failure, but reliability professionals strive to achieve proactive reliability.

STRATEGY

Proactive reliability is a strategy for preserving the reliability of machines or equipment. The reason for proactive reliability is to view machine failure and similar problems as something that can be foreseen and taken care of before they occur. It is a universal truth that it costs 10% of the failure's costs to be proactive and stop it from happening. Another way to look at this is to gauge the internal rate of return when comparing reducing your reliability costs over increasing sales. A 10% reduction in reliability costs in a plant creates a bottom-line effect equal to a 40% increase in sales. Which is easier, increasing sales by 40% or reducing reliability costs through proactive efforts by 10%? This can be seen in Figure 10.1. There is an interactive software file located at *www.deathofreliability.com.*

Financial Impact						
	Start		10% Maintanence		40% Sales	
	Dollars	Percent of Sales	Dollars	Percent of Sales	Dollars	Percent of Sales
Sales	100	100	100	100	140	100
Materials, Labor, & O/H	75	75	75	75	105	75
Maintenance Costs	20	20	18	18	28	20
Profit	5	5	7	7	7	5

Figure 10.1 Maintenance vs. Sales.

Proactive reliability concentrates primarily on identifying the root causes of machine failure and addressing those issues before problems occur. This is commonly referred to as *failure mode effects analysis* (FMEA). This approach is a systemic one and a key process used in proactive reliability. FMEA identifies how and where failures might occur, as well as what the impact of the failures would be so you can address the part of the process that needs attention. The review of the process required by FMEA includes: steps in the process, what can go wrong (failure modes), why will this happen (failure causes), and what the consequences are of each failure (failure effects). Organizations practicing proactive reliability use FMEA to review all possible failures and then prevent them by proactively correcting them instead of reacting to the failure events once they have occurred. By placing the emphasis on proactive elimination, you reduce the risks to people and equipment. This is useful in the design for reliability phase of proactive reliability because you can address the failure modes before they reach your plant floor.

Cost Effective

Proactive reliability is the most cost-effective practice since it allows a company to avoid machine failure and solves issues before they become problems. The proper use of proactive reliability helps avoid loss of productivity due to breakdowns or inoperable machines, and it saves a company a great deal of money.

Proactive reliability enacts actions directed at the sources of failure. It is designed to reach the design life of mechanical machinery as opposed to making repairs when nothing is broken, adopting failure as routine and normal, and preempting crisis failure maintenance.

For example, through research, it can be established that a certain machine is most likely to fail because of contaminants in the lubricating fluid. Once this is determined, then different proactive reliability methods can be used to guarantee that the fluid is clean to begin with and remains clean during the life cycle of the equipment. If contaminants are found within the lubricant, then it can be filtered out or replaced with a clean fluid that helps prevent machine breakdown. This stops the failure from ever happening and keeps the plant reliable.

A proactive reliability plan enables an organization to prolong the life of machinery and prevent a complete and unexpected breakdown of production. A proactive reliability plan lets an organization plan production shutdowns for repairs, inspection and maintenance.

While this type of proactive reliability may be confused with preventive measures, there are some clear differences between the two methods. Preventive maintenance typically relies upon proven conditions for potential breakdown or machine failure, and the use of regularly scheduled maintenance to avoid this. Proactive reliability gets out in front of the machine failure and attacks the cause before it becomes one.

What is "In"?

Major industries throughout the world have realized that it is time to throw out the old ideas on machine reliability. The cost-saving tendency is toward a reliability program that targets the root cause of equipment wear and failure. Predictive and preventive approaches are out: proactive reliability is in. But to achieve this we will need to open senior leaders' eyes about the need to hire a qualified reliability professional as the only path to sustainability.

Why? Because proactive reliability methods are presently saving industries of all sizes thousands, even millions of dollars in machine reliability every year. This concept of saving large amounts on reliability, however, may be tough for some to grasp.

In a study done by DuPont, maintenance was shown to be the single controllable expenditure in a plant. In many companies, it often surpasses annual net profit. The difficulty of costly maintenance has reached a serious level, but as some organizations have found out, and more come to comprehend every day, their maintenance costs can be cut considerably by beginning a "proactive" line of defense.

GETTING TO THE ROOT OF THE PROBLEM

When it comes to the life cycle of any machine, whether it's a lawn mower or a 1,000-horsepower bulldozer, keeping equipment clean, cool, and dry is key. Laboratory and field tests show that more than any other reason, fluid contamination is a top culprit of equipment failure; even the most minuscule particles can eventually grind a machine to a halt.

There are a number of accepted methods currently being employed to combat machine damage, and they are built on either discovering the warning signs of breakdown once they have already begun (predictive) or component replacement per a schedule rather than the machine's true condition (preventive).

Recently, a new discipline has begun to take a micro view on machine damage, focused on the cause instead of the symptoms of wear. Proactive reliability is that discipline, and it is being acknowledged worldwide as the most important means of achieving savings, unsurpassed by conventional maintenance approaches.

In a proactive reliability strategy, you monitor the correction of failing root causes: improper lubrication, contamination, or improper installation. This is like managing your blood pressure and cholesterol with a proper diet. In a predictive maintenance strategy, you monitor equipment or components that have begun to fail because of wear, debris, heat, vibration, or alignment. This is like evaluating the damage to your heart with an electrocardiogram (EKG) or ultrasound. In a preventive maintenance strategy, you replace components at a set frequency. This is like addressing issues with your heart with a bypass or a transplant. The final maintenance strategy is run-to-failure; for this you need a large maintenance budget, and it is like having a heart attack or stroke and then doing what it takes to recover. Usually, it involves damage to other components as well as the complete failure of the component in question. It is easy to see which strategy makes the most sense. It costs 10% of the total breakdown costs to prevent the breakdown with a proactive strategy.

PROACTIVE VS. PREVENTIVE/PREDICTIVE

Imagine being able to identify and eliminate a disease before any symptoms occur in your body. It would save you money and prevent you from having to stay in the hospital in the long run. This is the benefit of proactive reliability over all others.

Proactive reliability dictates corrective actions pointed at the sources of failure. It is created to increase the life of mechanical machinery as contrasting to: (1) effecting repairs when often nothing is broken, (2) accommodating failure as usual and normal, and (3) preempting crisis failure maintenance, all of which are characteristics of the predictive/preventive disciplines.

While effective to a degree, neither preventive nor predictive maintenance is geared to delete any of the three common and serious failure culprits: improper lubrication, contamination, or improper installation.

There are a number of methods that are used to enact a proactive reliability solution. For instance, making sure to train workers in the right procedures for machine operation or using a group of reliable suppliers for machine systems like lubrication and parts are excellent ways to help enhance reliability systems.

A BALANCED APPROACH

Ultimately, the top machine reliability plan is one that balances fundamentals of each of these modes. The Department of Energy's Office of Energy Efficiency and Renewable Energy (EERE) documents that reliability-centered maintenance programs are found throughout the top performing industrial facilities in the country. These program costs are mostly less than 10% run-to-failure, between 25 and 35% preventive and between 45 and 55% predictive.

The goal of a reliability program is to make sure that the facility is constantly up and running. Thus, it is crucial to ensure that these plans include the machines that are unique to a plant and its operations. Now let's look at the three reasons we have unreliability; proactive reliability addresses each of these before they make your operation unreliable. By properly storing, handling, delivering, and managing your lubricants you will proactively stop failure related to improper lubrication. By implementing and managing a contamination control program that includes lubricants and spare parts, you will proactively stop failures related to contamination. Finally, by ensuring proper installation (which includes developing your operators and reliability teams, managing your spare parts inventory, and designing your plant equipment for reliability), you will proactively stop failures related to improper installation.

CHAPTER 11

CONCLUSION

Having read all of this, where do you go from here? It is all about the simple common sense approach. Eliminate the three reasons organizations suffer from unreliability: improper lubrication, contamination, and improper installation. I have chronicled steps you can follow to educate and move your organization through these. I have repeatedly mentioned that there is no shortcut, easy path, or quick fix. Real-world reliability takes blood, sweat, and tears. The leadership of the teams tasked with this accomplishment must be true reliability professionals, not engineers, production personnel, or consultants. If you are looking at these folks to drive this work, you will fail.

I have heard a lot of reasons as to why these shortcuts and quick fixes work. I would challenge anyone trying to sell an organization on this false belief. These same people mask issues and lie about results. They may convince the organization that they are successful, but eventually, someone will ask the right question or question why their production has not increased or even maintained its average throughput. They will begin to question how to become more competitive, where to gain competitive advantage, or how to curb escalating maintenance costs without a corresponding increase in production. At this moment, many of these organizations will make the costly mistake of hiring a consultant. I have discussed how to use and how not to use consultants in your organization. If you are looking for a consultant to rescue you, you will die. Consultants are not invested in your success, but in their bottom line. If you think one of your engineers can manage it, you will fail. If you think your best production manager can handle it, you might as well let the maintenance team lead itself. Taking shortcuts with respect to reliability is like cutting your own throat. Stop doing it!

REASON FOR FAILURE

Leaders in organizations who think like this are the single largest reason maintenance and reliability are failing. It is the main cause of the loss of skilled tradespeople. Bad decisions masked as lean manufacturing and other misguided initiatives that lack the real vision necessary to sustain competitive operations are other drivers. Parents, teachers, and our government contribute to this issue by misleading our young people into thinking college is for everyone and the answer to their lives. The attitude of young people is the least of the problem, other than what has been beaten into them by parent and teachers.

I want to again state that my condemnation of unqualified people leading reliability is not directed at the individuals struggling to fill these positions, but squarely at the senior leaders making unqualified, shortsighted, short-term decisions because they themselves are not truly qualified for their jobs. Yes, I said they are *not* qualified, and to see this all you have to do is look at this subject and the loss of the leadership on the world stage.

HIRING THE RIGHT LEADER

Hiring the right leaders is the key to success and I am always amused, but frightened, by the fact that most organizations charge the selection and development of their leaders to their Human Resources (HR) Departments. These departments are led by individuals who have never lead a team or worked on the plant floor. They have limited education and no practical experience in leadership. Most of them feel that training and development are the same thing, and this is wrong. You cannot train leaders, you have to develop them and that only happens in the field and on the floor with their teams. An HR generalist or manager cannot lead this development and lacks the ability and experience to pull this off. But like all of the other failed approaches I have discussed in this book, HR personnel are unwilling to acknowledge that they do not possess the necessary skills for fear of losing their jobs. How can they impart this knowledge to those requiring it? Fear of the perception that they are not doing their jobs drives their unwillingness to act in the best interest of their organizations. There is a gap between theory and reality, and this reality led me to pursue my doctorate in order to be qualified to develop leaders and have the practical experience to help them reflect on real-world application in leadership.

Another aspect to leadership is the ability to maximize your resources. Companies boast that they have removed the walls in their organizations to create an environment of teamwork. I often look to the lean elimination of waste. One of these wastes is the waste of skill. Underutilized capabilities manifest themselves in a lot of ways. The assumption that the person holding a specific position in an organization is the plant's subject matter expert can and often does lead to this waste. I struggle in my jobs with having to tiptoe around HR managers when it comes to developing leaders. In my experience, I have yet to meet a HR manager who has the education or experience I have in developing leaders. The focus of my doctoral research and all subsequent research and work has been focused on leadership development. I actively participate in upper echelon and leadership research as well as conduct webinars and keynote speeches on this subject. My background in this area is far beyond theory and yet I have to bite my tongue when confronted by what I know are outdated and failed approaches to developing leaders. If an organization only looks to the person holding a title for expertise, they are probably wasting a lot of skills.

HOW DO WE TURN THE TIDE?

The answer is one organization and one leader at a time. You need to start by collecting some basic data. What is your plant's capacity? What is your plant's efficiency? This is probably an established metric like *overall equipment efficiency* (OEE), or you can approximate it by taking the actual production divided by the theoretical production if you ran at maximum speed for all your operating hours.

Next, understand what the capital costs are for adding new equipment. If you have one production line, what would it cost to add a second? If your company has three factories, what would it cost to add a fourth? What is the operating cost in labor and utilities to run that new line or new plant?

Now it is time to do some math! If you are currently running at 50% efficiency and you believe you can improve that by 10 points to 60% with "more reliability," that will result in a whopping 20% gain in capacity!

(60%-50%) / 50% X 100 = 20%

The figures work out the same if you are starting at 40%, but now a 10-point improvement increases capacity by 25%. To determine the opportunity to increase throughput, I have provided interactive software located at *www.deathofreliability.com.*

Would your organization invest $200,000 or some other amount annually in "reliability" to get this level of capacity increase as an additional line that will cost a million dollars or more to install and $500,000 annually to staff and run?

When you swap your own estimates, the resulting assessment will be different in the details, but it will always favor an approach that improves reliability. This is notably true for a growing organization that is investing in machinery to prevent being capacity inhibited.

What if you are not growth or capacity constrained? The equivalent reliability enhancements proven above will reduce your operating labor, utilities, and machinery wear and tear by 18 to 25%. Put that in your Return on Investment calculator.

Let me close with just a few salient points:

- Many organizations with unstructured maintenance programs run as poorly as 40% efficiency.
- If you run above 80% efficiency, the math still works, but the gains are less, just like adding additional insulation to an already properly insulated attic.
- It is easy to improve a poorly running operation by 10 points. I have inherited facilities running at 30% uptime, and had improved it to 85% in three years using many of the proactive approaches described throughout this book. This almost tripled our capacity, and we had room for even more improvement. You already have some ideas about how to enhance your place.

As a qualified reliability professional, you have to make some estimates and develop plans. Ultimately, you will need to sell your proposition and be accountable for the results if an "investment" is to be made. This is the same as for a project manager who is building a new plant or line. Organizing your project plans into a one–and five–year plan will help you focus your efforts and strengthen your budgeting ability. I have included a one– and five–year plan at *www.deathofreliability.com* to help you with this. I use the P-F Curve to illustrate to senior leaders the impact of proactive reliability.

P-F CURVE

When trying to understand the relationship between the four maintenance strategies, I use the P-F Curve to illustrate it. The P-F Curve was developed in 1978 by a study and ultimately a report generated by United Airlines but funded by the U.S. Department of Defense. It was developed as a guideline for reliability efforts. The sad thing is, if you ask 100 maintenance and reliability professionals what the P-F Curve is, 10% will probably know. Of that 10%, normally less than one percent will really understand it. The intent of the P-F Curve is to graphically show leaders in industry how proactive reliability can improve plant performance.

To explain it simply, the *P* in the P-F Curve stands for the point where the functional failure can be identified and measured and the *F* stands for the point in an equipment's life cycle where functional failure is imminent. A plant's ability to identify these points is dependent on three things: a definition of the function of the item, what constitutes failure, and what indicates when failure will occur. About 90% of equipment fails randomly. It is determined by many sources, and interval or time-based maintenance will not effectively address plant reliability. To help you visualize the curve and its interconnection with maintenance strategies, see Figure 11.1.

In the 1960's, organizations employed preventive or interval maintenance because they focused on guessing the age something would fail and not why it was failing.

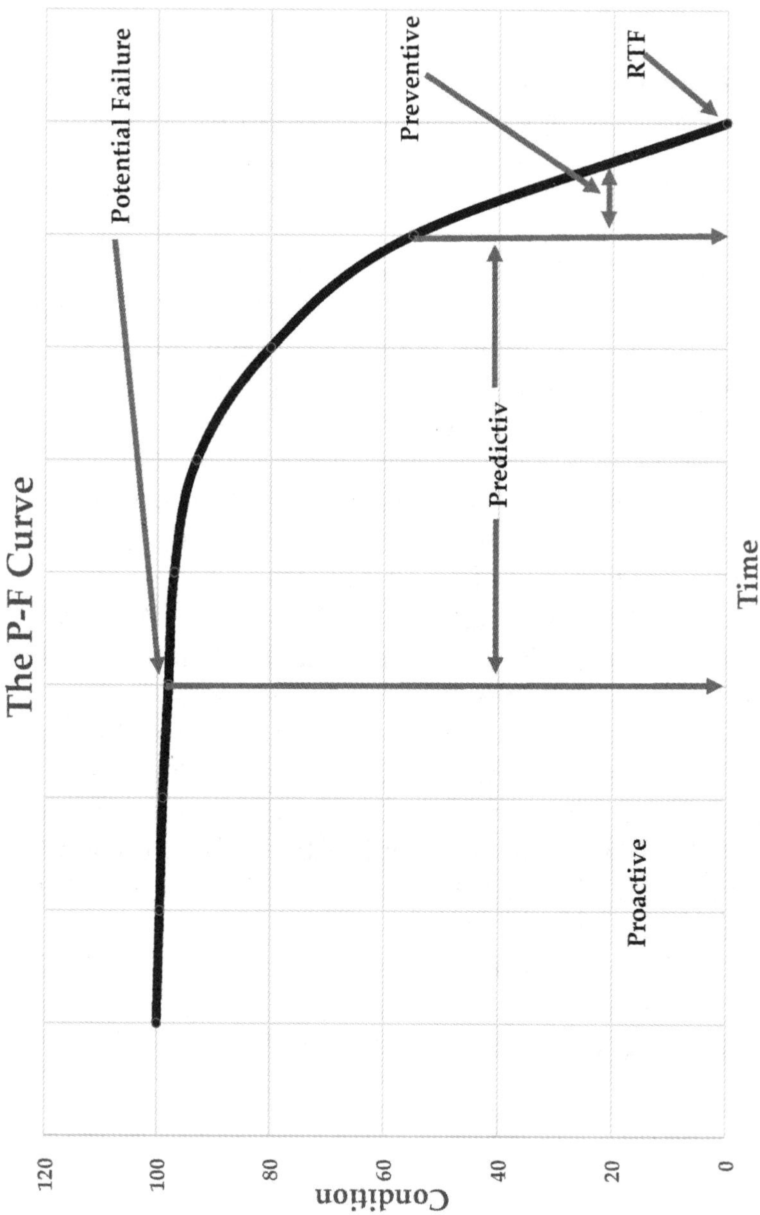

Figure 11.1 The P-F Curve.

Leaders Do Not Understand the P-F Curve

The majority of leaders do not understand how the P-F Curve relates to the four maintenance strategies. I am asked all the time what the difference is between reactive and proactive reliability. Most non-maintenance and reliability professionals view proactive reliability as anything that identifies work in advance. In their definition, three of the four maintenance strategies would be considered proactive. I disagree because all but *true* proactive reliability is reacting to a failure indication and not eliminating the failure. Preventive maintenance is interval-based and that interval is based on the anticipated failure of the equipment. This is reactive. Predictive maintenance monitors and predicts the failure of the equipment. Vibration, thermal, and others require the equipment to begin to fail before it starts to predict it. The proactive measures are proper lubrication, contamination control, and proper installation. It is that simple.

Organizations often do not understand that just because a machine is running does not mean it is not failing. The use of the P-F Curve can assist in communicating the importance of the relationship between cost and breakdown and how we can save money. The earliest warning signs of equipment failure cannot be detected, so proactive reliability is the only way to stop premature failures.

Equipment will run even after it has begun to fail. But once it has begun to fail it is only a matter of time before it does so completely. Depending on the maintenance approach, the cost of repair is dependent upon how early you identify it and address it. It saves your organization up to 90% of the failure costs to address it proactively. Who could not use a 90% savings? When asked where we will get the money to be proactive, the answer is by not letting it fail. The earlier we get out in front, the larger the savings. Preventive efforts save you 12 to 18% over run-to-failure, predicative efforts can save up to 50%, and proactive can save 90%. There is your cost justification.

WILL IT RUN FOREVER?

Organizations need to understand that equipment is not designed to run forever, but we can help it live a longer and happier life. It has grown increasingly more difficult to convince senior leaders to do the right thing when it comes to their equipment. Proactive reliability is not a hassle or expense, but the right thing to do for organizations with the focus on long-term success. The erosion of true pedigreed managers is leading to the death of reliability, and those of us left to combat this death need to act now.

Just think if you had more resources; what would you do, how much would it cost you, and how much do you think you could increase your throughput? Lastly, how does this compare with the capital cost of adding a similar amount of capacity?

Speaking of costs, I always get asked, "How do you fund all of this?" The answer is quite simple: invest in reliability. When I start the pursuit of reliability, I start with proper lubrication. This has the biggest bang for your buck. Most organizations can capitalize the necessities to develop and implement a proper lubrication program. From this you will realize significant savings. Here is the tough part; you need to convince your senior leaders to reinvest this savings back into the next phase of implementation of proactive reliability. Do not pull the savings from the reliability budget but use it to pay for the next improvement.

Continue this approach until you have achieved a proactive reliability strategy. At this point, you take the savings out of the reliability budget. It is a simple solution to the question of where you will get the money.

Reliability can only be achieved through hard work and equity sweat. There are no shortcuts or quick fixes. If you are looking for immediate satisfaction, you will be disappointed. It must be led by a qualified reliability professional, as defined above. Here is your toughest challenge, because qualified reliability professionals are a dying breed. It is in direct relation to the dying reliability across all industries. You do not need to look any further than this fact and this correlation to see that everything I have said in this book is true. We are facing the "Death of Reliability" if we do nothing. I am calling upon what is left of the qualified reliability professionals to step up and stand with me to revive reliability and the skilled tradespeople necessary to rehabilitate our profession.

Recommended Reading

Avolio, B. J. (2005). *Leadership Development in Balance: Made/Born.* New York, NY; Lawrence Erlbaum Associates, Inc.

Collins, J. (2001). *Good to Great Why Some Companies Make the Leap… and Others Don't.* New York, NY; HarperCollins Publisher, Inc.

Daley, D.T. (2008). *The Little Black Book of Reliability Management.* New York, NY; Industrial Press, Inc.

Deaton, D. R., (2009). *Ownership Spirit - The One Grand Key That Changes Everything Else.* Mesa, AZ: Quma Learning Systems, Inc.

References

Avolio, B. J. (2005). *Leadership Development in Balance: Made/Born.* New York, NY; Lawrence Erlbaum Associates, Inc.

Avolio, B. J., Bass, B. M., & Jung, D. (1999). Re-examing the components of transformational and transactional leadership using the Multifactor Leadership Questionnaire. *Journal of Occupational & Organizational Psychology,* 72, 441-462

Bass, B. M. (1998). *Transformational leadership: Industrial, military, and educational impact.* Mahwah, NJ: Lawrence Erlbaum Associates, Inc.

Bass, B. M. (1999). Two decades of research and development in transformational leadership. *European Journal of Work & Organizational Psychology,* 8, 9-33.

Collins, J. (2001). *Good to Great Why Some Companies Make the Leap ... and Others Don't.* New York, NY; HarperCollins Publisher, Inc.

Conger, J. A, & Benjamin, B. (1999). Building leaders: *How successful companies develop the next generation.* San Francisco: Jossey-Bass.

Daley, D.T. (2008). *The Little Black Book of Reliability Management.* New York, NY; Industrial Press, Inc.

Deaton, D. R., (2009). *Ownership Spirit - The One Grand Key That Changes Everything Else.* Mesa, AZ: Quma Learning Systems, Inc.

Offermann, L. R., Hanges, P. J., & Day, D. V. (2001). Leaders, followers, and values: progress and prospects for theory and research. *Leadership Quarterly,* 12, 129-132.

Seibert, K. W. (1999). Reflection-in-action: Tools for cultivating on-the-job learning conditions. *Organizational Dynamics,* 28, 54-65..

INDEX